## Advance Praise for
## *Sick*

"Porochista Khakpour's powerful memoir, *Sick*, reads like a mystery and a reckoning with a love song at its core. Humane, searching, and unapologetic, *Sick* is about the thin lines and vast distances between illness and wellness, healing and suffering, the body and the self. Khakpour takes us all the way in on her struggle toward health with an intelligence and intimacy that moved, informed, and astonished me."

—Cheryl Strayed, *New York Times* bestselling author of *Wild*

"Sickness, in the world and in the body, is an attempt to flatten the individual, to make it conform to an inflexible name. Porochista Khakpour resists this on every page. Her writing is first of all vibrant, humming, strong, tall, striding. It powers through paper frailties. Survival, she reminds us at the end of *Sick*, can be an act of the imagination: it is the courage to insist on seeing yourself decades in the future, climbing a mountain, squinting into the sun, sitting down at the desk to write what happened."

—Patricia Lockwood, author of *Priestdaddy*, named one of the 10 best books of 2017 by the *New York Times*

"Porochista Khakpour's *Sick* is a bruising reminder and subtle revelation that the lines between a sick human being and a sick nation are often not lines at all. The book boldly asserts that a nation wholly disinterested in what really constitutes 'health' will never tend the bodily and emotional needs of its sick and vulnerable. Somehow, Khakpour manages to craft the minutiae of the moments spent keeping herself alive while obliterating what could have easily been written as spectacular melodrama. I'm most amazed at how time itself, and point of view, are 'sick' and 'sickening' in this wonderful memoir. Khakpour has done more than something I've never seen before in this phenomenal book; she's done something I never imagined possible."

—Kiese Laymon, author of *Heavy*

"I'm so excited for the world (you!) to read Porochista Khakpour's *Sick* because now you'll understand. Understand what it's like to navigate a broken medical system; understand what chronic illness does to the self; understand the damage that doubt and ignorance can wreak; understand how living and self-destructing, writing and working, loving and sex doesn't just stop when you're ill. And for those of you who understand this all too well, this book gives a voice—a fierce, booming, brutally honest voice—to the millions of people silently suffering with invisible illnesses of their own. 'I always felt broken in my body,' she writes, and I shudder with recognition. Thank you, thank you, thank you, Porochista, for giving so much of yourself in this miraculous memoir. The world is a better place with your book in it."

—Susannah Cahalan, #1 *New York Times* bestselling author of *Brain on Fire*

"Thank you, Porochista Khakpour, for writing an unflinchingly honest, complicated memoir about living life with Lyme. *Sick* should be required reading at every medical school!" —Kathleen Hanna

"*Sick* is a riveting plunge into the most profound mysteries of mind and body—the haunted labyrinths of addiction; a chronic illness that mightily resists answers; and, ultimately, a diagnosis that proves just as confounding: late-stage Lyme disease. As Porochista Khakpour works to uncover the roots of the maladies upending her physical and mental health, she raises vital questions that challenge the common perceptions around illness and treatment and recovery. Miraculously, *Sick* emerges as a force of life."

—Laura van den Berg, author of *The Third Hotel*

"This is a story of towering frustrations written so beautifully that through the weird alchemy of art it ends up lifting the reader's spirits. You read these elegant sentences and get the elusive click that you get in the presence of the real thing. To the list of brilliant fiction writers penning timeless memoirs—Nabokov, McBride, Wright, Styron, Ward, Gay, both Woolfs, to name a few—we now indelibly add the name Khakpour. Khakpour battles a disease that attacks the quality of one's life in every way, but perhaps the most poignant element here is the world's lack of faith in her affliction, so that she faces the double indignity of fighting a fearsome foe on the one hand and arguing for its potency on the other. This is a gripping, moving, thoughtful meditation written at the highest levels of narrative engagement."

—Matthew Thomas, *New York Times* bestselling author of *We Are Not Ourselves*

"This is a book that throws me into the time of my own being. I experience Porochista Khakpour's *Sick* as an act of radical friendship because nobody should know this much about anybody else unless they love each other, and this book, so quotable and well-phrased at absolutely the worst of moments, and there is a lot of 'worst' here—because this is a book of physical suffering—is stalwartly framed by love: of family and friends and sex and all kinds of partnership as the activist bedrock of health, and finally love of the city too. Born in Tehran, Iranian American author Porochista Khakpour habitually picks New York City as her sanity and her chosen rite of return. Thrumming, diaristic, unabashedly wild, and homeless-feeling, *Sick* is something gut-wrenching and new, a globally intimate book."

—Eileen Myles, author of *Afterglow*

"This memoir is not your traditional illness narrative. Porochista Khakpour threads together a startling tapestry of stories about a young woman seeking place—in the America she flees to as a refugee of Iran, in a medical system that offers her no answers, in the empty promises of pill bottles and dangerous lovers, and ultimately, in the body. Electric, daring, and staggeringly honest, Khakpour's writing takes us to the very edges of what it means to be alive."

—Suleika Jaouad, author of the *New York Times* "Life, Interrupted" column and video series

"*Sick* stages on the page what is at stake for a body under endless siege from addiction, illness, trauma, dislocation and dispossession. The questions emerging from this body story challenge ideas about identity and the too-easy logic of sickness and health, as well as the bicultural boundaries of being. What does it mean to be alive inside a raging body? By sharing her body story, Porochista Khakpour gives the reader a profoundly generous gift: an unflinching narrative of the deep desire to live. *Sick* is a triumph of the imagination as she holds her heart out to you."

—Lidia Yuknavitch, national bestselling author of *The Book of Joan*

"Readers, writers, and sick people all crave origin stories, and Khakpour tries to serve one up: the first bad decision, the first bad boyfriend, the first childhood trauma? A tick bite in California? Pennsylvania? New York? Which came first, addiction or infection? Then she shows us the oceanic mess that is chronic illness, a story without a clear origin or a neat arc, and we see how it becomes an ongoing presence—not a narrative at all. We are forced to consider what it must be like to live like this, to leave aside all our illusions of fairness, logic, and control. What a gift."

—Sarah Manguso, author of *300 Arguments*

ALSO BY POROCHISTA KHAKPOUR

*Sons and Other Flammable Objects*
*The Last Illusion*

# SICK

*A MEMOIR*

**POROCHISTA KHAKPOUR**

HARPER PERENNIAL

NEW YORK • LONDON • TORONTO • SYDNEY • NEW DELHI • AUCKLAND

HARPER PERENNIAL

The names and identifying characteristics of some of the individuals featured throughout this book have been changed to protect their privacy.



HarperCollins books may be purchased for educational, business, or sales promotional use. For information, please email the Special Markets Department at SPsales@harpercollins.com.

FIRST EDITION

*Designed by Leydiana Rodriguez*

*Title page image of pills © Robert Kneschke / Shutterstock, Inc.*

Library of Congress Cataloging-in-Publication Data has been applied for.

ISBN 978-0-06-242873-8

18 19 20 21 22 LSC 10 9 8 7 6 5 4 3 2 1

This book contains my personal story. I am not a medical professional, and, therefore, the inadvertent advice and information I share throughout this book is in no way intended to be construed as medical advice. If you know or suspect that you have a health problem, it is recommended that you seek the advice of your physician or other professional advisor before embarking on any medical program or treatment.

*To Voyce*
*& her honeybees*

"Those great wars which the body wages with the mind a slave to it, in the solitude of the bedroom against the assault of fever or the oncome of melancholia, are neglected. Nor is the reason far to seek. To look these things squarely in the face would need the courage of a lion tamer; a robust philosophy; a reason rooted in the bowels of the earth."

—VIRGINIA WOOLF, *On Being Ill*

"Do you believe, she went on, that the past dies?
Yes, said Margaret. Yes, if the present cuts its throat."

—LEONORA CARRINGTON, *The Seventh Horse and Other Tales*

# SICK

# AUTHOR'S NOTE

It seems impossible to tell this story without getting the few certainties out of the way, the closest one can come to "facts." The hardest part of living with Lyme disease for me has always been the lack of concrete "knowns" and how much they tend to morph and blur over the years, with the medical community and public perception and even within my own body. To pinpoint this disease, to define it, in and of itself is something of a labor already.

Still: Lyme disease is a clinical diagnosis, a disease that is transmitted by a tick bite. The disease is caused by a spiral-shaped bacteria (spirochete) called *Borrelia burgdorferi.* The Lyme spirochete can cause infection of multiple organs and produce a wide range of symptoms. Less than half of Lyme patients recall seeing a tick bite, and less than half also report seeing any rash. (They say the deer tick—which is usually the carrier of Lyme—can present as smaller than a speck of pepper.) The erythema migrans (EM) or "bull's-eye" rash is considered the main sign of Lyme, but atypical forms of this rash are seen more frequently. Testing is quite flawed; the commonly used

ELISA (enzyme-linked immunosorbent assay) screening test is unreliable, missing 35 percent of culture-proven Lyme disease. There are five subspecies of *B. burgdorferi,* over one hundred strains in the USA, and three hundred strains worldwide. Testing for babesia, anaplasma, ehrlichia, and bartonella (other tick-transmitted organisms) should always be performed as well, as coinfection with these organisms points to probable infection with Lyme and vice versa.

There are multiple stages and progressions of the disease. Stage 1 is called early localized Lyme disease, and it signifies a stage where the bacteria have not yet spread throughout the body; this form of Lyme can be cured with timely antibiotic use. Stage 2 is called early disseminated Lyme disease, and here the bacteria have begun to spread throughout the body. Stages 3 and 4 are often known as chronic and late-stage Lyme disease, and at this point the bacteria have spread throughout the body. Many patients with chronic Lyme disease require prolonged treatment, all while relapses may occur and retreatment may be required. There are no tests to prove that the organism is at any point eradicated or that the patient with chronic Lyme disease is "cured," although one can test for inflammation and other markers. Each year, approximately thirty thousand cases of Lyme disease are reported to the CDC. Over the past sixty years, the number of new cases per decade has almost quadrupled; the number of outbreaks each year has more than tripled since 1980.

I have Lyme, with "bands" (lines on a test that represent antibodies to different components of the bacteria) that afford it CDC-level recognition (bands 23 and 41). My main coinfection has been ehrlichia. Several doctors believe I also have babesia and bartonella due to certain symptoms, although my tests don't always come out positive for them.

Living with this disease has cost me more than $140,000 so far. Experts put the average cost of late-stage Lyme at somewhere around $20,000 to $200,000. The annual cost of Lyme disease in the United States is more than $1–$3 billion as of 2017.

It is unclear when I got the disease. Doctors have mostly pinpointed somewhere in the 2006 to 2009 range, but I've had doctors who think I've had it since childhood. Although the disease and its complications—including addictions—have defined my life, it is unlikely I will ever know when I contracted it, just as it is unlikely I will ever be rid of it entirely.

# ON THE WRONG BODY

I have never been comfortable in my own body. Rather, I've felt my whole life that I was born in the wrong body. A slight woman, femme in appearance, olive skin that has varied from dark to light, thick black curly hair, large eyes, hands and feet too big, of somewhat more than average height and somewhat less than average weight—I've tried my whole life to understand what it is that seems off to me. It's deeper than gender and sexuality, more complicated than just surface appearances. Sometimes the dysmorphia I experience in my body feels purely psychological and other times it feels like something weirder. As a child, I thought of myself as a ghost, an essence at best who'd entered some incorrect form. As I grew older, I accepted it as "otherness," a feature of Americanness even. But every room I walk into I still quickly assign myself to outsider status, though it seems not everyone can see this. Many have in fact called my looks conventional, normal, even "good." I've accepted it while also feeling like I've deceived them.

I've looked for answers from my first few years on this earth, early PTSD upon PTSD, marked by revolution and then war and then refugee years, a person without a home. Could that have caused it? Was displacement of the body literally causing a feeling of displacement in the body?

Only decades later did I confront something that may have been there the whole time: illness, or some failure of the physical body due to something outside of me, that I did not create, that my parents did not create. Illness taught me that something was wrong, more wrong than being born or living in the wrong place. My body never felt at ease; it was perhaps battling something before I knew it was. It was trying to get me out of something I could not imagine.

At some point, with chronic illness and disability, I grew to feel at home. My body was wrong, and through data, we could prove that.

Because my illness at this stage has no cure, I can forever own this discomfort of the body. I can always say this was all a mistake. To find a home in my body is to tell a story that doesn't exist. I am a foreigner, but in ways that go much deeper than I thought, under the epidermis and into the blood cells. I have started to consider that I will never be at home, perhaps not even in death.

# PROLOGUE

It's New Year's Eve, about to turn 2016, and I've been where I always am: inside. A neighbor visits and drops off some left-over Christmas chocolate I can't eat but gladly accept; a friend a few blocks away comes by with his toddler son and invites me to his home for a small party that we both know I can't attend; friends all over the city send email and text invites to events "just in case." I've never not been a party girl. This was my father's greatest fear for me in the United States, but one that I balanced with what would become his greatest dream for me: being an author. *A cross between Salman Rushdie and Paris Hilton*, he used to joke.

But I know that this New Year's there will be no parties for me. This New Year's will be my first spent alone.

Twenty days before, I was in a car accident. Hit by a semi—an eighteen-wheeler tractor-trailer, to be precise—on the way home from my job, teaching at Bard College. Class was out by 1:10, but I stayed late that Friday, rare for me. It had been a season of hate crimes, a month and a half after the Paris attacks, and tensions were especially sky-high for brown and black students. I was the only faculty of color in our department that term and the students seemed to look to me for answers. I hid that I was as lost as them. I had packed extra snacks and all sorts of "reinforcements," as I'd call the supplements that I'd been taking for years for Lyme disease relapses—from Celtic sea salt to magnesium to nuts to protein shake mixes to bee pollen and propolis. These reinforcements were meant to shore me up, so that I could stay a few extra hours and meet with all my students who seemed to have some sort of depression that season.

I understood: so did I.

---

The first sign of a Lyme relapse is always psychiatric for me. First the thick burnt fog of melancholy that crept slowly—mornings when I couldn't quite get out of bed, sticky inability to express my thoughts, hot pangs of fear and cold dread at unpredictable times, a foundation of anxiety, and panic—that fluorescent spiked thing, all energy gone bad, attacking like clockwork around noon daily—all unified toward that endless evil white, insomnia.

Everything was again a danger, everywhere and everyone and every time.

Days after I returned from a blissful but exhausting book festival in Indonesia that November, I began to consider that I might be having a Lyme relapse. At that point, I'd been healthy for years, so to relapse into bad health was a transition I couldn't quite fathom. I tried to blame jet lag at first, the disappointment of leaving a wonderful place like Indonesia, the hectic schedule they had me on (three Indonesian cities, all spread apart, in ten days). I tried to think it might be the news I got upon return: that my editor was leaving publishing, and that maybe this very book would be in jeopardy. I tried to think it was the Paris attacks and the new wave of Islamophobia that had suddenly gone mainstream. I tried to think anything, everything else.

I wasn't going to lose myself again.

After a Thanksgiving spent intentionally alone—I never liked people seeing me in an off-period—where the one event of my holiday was finally caving in to my doctor's suggestions and buying a cane from the local CVS, I broke down and wrote my friends an email on November 28.

> dear some of my closest nyc friends who are in town currently or might be soon,
>
> i am getting more and more ill very fast. i'm scared at the moment. in case you don't know, i've had a late stage lyme relapse but this one feels very intense. rapidly things are going downhill.
>
> i'm trying not to be extra alarming online—some important work stuff i want to be well enough for—while also letting people know some things are off.
>
> i have various lyme communities and that's the way to reach them.
>
> but also i don't want to drop out as last time i became completely disabled that way—i need to stay engaged
>
> but i'm scared.
>
> at points in the day i don't know where i am exactly. at night it somewhat clears.
>
> i've been falling again a lot. etc. very faint, very dizzy. getting a cane.
>
> having trouble with reading and writing.

it's very reassuring to be around people when I'm confused. alone it is very hard.

i'm not totally sure what i'm asking.

it's hard to ask for help here because what can you do even? i don't have the imagination to know what is help right now completely

but some things maybe

would you mind occasionally checking in on me? i might not be able to text back very effusively as i'd wish but perhaps briefly—it doesn't mean i am mad, it just means i can't think. i'm also worried something will happen to me and cosmo will just be alone

also if anyone had the time and was interested in being a passenger in my car with me? i have 2.5 more weeks of classes.

for now i think i can drive. or was able last week. but i go to bard twice a week (weds and thurs) and sarah lawrence mon and tues.

i have nice offices at both if you wanted to work there too and even hang out with cosmo?

or if you were ever in harlem? just walking with me to the dogpark? i'm not that deep in harlem, just 120th.

or perhaps riding the subway with me? (i tend to avoid subways alone when relapsing)

i might even drive cross country if i find i can't take cosmo with me on a plane to the west during break. just will do the southern route slowly. if anyone wanted a free ride there too!

i'm also happy to pay anyone for their time. not meant to be insulting! just meant to say i value your time.

basically just the presence of others around me right now is helpful i think.

(tho i also have a lot of work to do so i can't take breaks. perhaps study dates?)

all my friends are busy people who do work i love so i'm hesitant to ask. also you all have your own shit right now.

i will not be mad at all if you can't deal with this right now! i'm embarrassed to ask frankly. i thought to write people individually but i didn't want you to feel the burden like you were the only one!!

basically i am very bad at this.

and sorry for chaotic nature of this email. hard to express myself.

love p

Most of my friends had never received an email like that from me. When I had my first definitive Lyme Crisis—what I now call 2011–2013—I removed myself from many people's lives, while some removed themselves from mine. Here and there friends stuck by me, sometimes a partner, but the only consistent presence was a few doctors. Then, my dog, Cosmo. I was by no means alone—I had distant but steady support, but all in all, when it came down to it, it was me alone going into it and me alone coming out of it: driving cross-country in the dead of winter just past remission to pick up life somewhere, anywhere.

And here I was again.

My doctor did what he did the past two times I had had

Lyme relapses and prescribed me supplements and medications: words like *glutathione*, *acetylcholine*, *methylfolate*, *fluconazole*, and more all back in my life. I was back to having dozens and dozens of pill bottles. It was done in an email and a phone call. And it all felt more or less under control for some weeks.

Still, I was cautious and did not drive the two hours back and forth to Bard College, where just the year before I'd been appointed "writer in residence," and to Sarah Lawrence, where I'd been adjuncting to make the Bard job financially feasible.

In that penultimate week of the semester, I had a bad experience with my usual cab ride from Bard to the Poughkeepsie station, where I'd catch the Metro North to my apartment in Harlem. My forty-minute cab ride to the station was usually uneventful, but this time my driver Alan confessed to me that he was back on drugs. We had spoken about drugs before, on another ride I had taken with him, so he felt comfortable getting into it with me.

"You know how it is, Porshka," he said. He could never say my name. "Come on, you know how it is."

"I do know," I said cautiously, as I noticed him speeding faster and faster. "And that's why I think you should be . . . careful."

I was tempted to ask him if he was on drugs right then and there, but I had my answer, I thought. I tried to make out his speedometer.

"You know, I gotta make some money on the side," he said absently. He had recently begun dating a woman who worked in real estate. He had mentioned they "have a lot of fun. She likes it when I share my coke."

I had tried to switch the subject to the weather, to my students, to all the generic things we used to complain about together. (I tried to avoid talking about the Middle East—too many times I could tell Alan and I were not politically aligned and I didn't want to push the subject.) I even got to Lyme. The ride was nearly an hour, always so much space to fill, but this time felt particularly taxing. I tried to concentrate on the trees blurring by: maple, oak, hemlock, cedar, pine, all still lush in that season of little snow, starkly stabbing into the immaculate blue of twilight. I wanted him to slow down, but I also wanted the ride to end.

"You take anything for it?" he was asking.

I realized I was barely listening to his end of things.

"The Lyme—you take any pills? You need pills for it? Pain pills?" he was asking.

I remembered he had once told me his mother was addicted to some "pain pill."

"I don't take those kind of meds," I said. "I take mostly supplements. Nutrients. I don't do that other stuff."

"Anymore," Alan said with a smile, as he pulled in to the station.

"Anymore," I decided to agree with him—he was not wrong, after all—but I also decided for myself that there would be no more Alan in my near future, that the bulk of the semester away from him had actually been good for me.

The Poughkeepsie train station was less bleak than usual that night, the holidays in the air. As I went up to the ticket counter, I realized it was the same attendant I always used to see, a guy I nicknamed Lou because he looked like a Lou, while really his name was something like Lawrence. He would always ask me out at the end of our transaction, often a *I hope you don't mind me asking but you got a boyfriend?* I always said I did even though I never did during my time commuting. He'd always tag on, *Well, keep me in mind*. I'd always throw him something between a nod and a shrug and walk off with a weak wave.

This time he looked shocked to see me.

"Yeah, it's been a while!" I said quickly, hoping this time I could avoid his propositions.

"No, no, look at you, what happened?" he cried. He pointed to my cane. "Why do you have . . . *that*?"

I waved off his question as I often did. "Lyme. I get dizzy, that's all."

He kept staring at the cane, and then I realized his eyes were welling up with tears.

I nearly laughed it was so absurd. "Hey, are you okay? It's nothing, I'm fine. I've dealt with this for years!"

He shook his head. "No, no, it's just that . . . Lyme." I should have remembered that, unlike in the city, upstate everyone knew about the severity of Lyme disease. But I wasn't ready for what he said next: "My father passed away a couple months ago from complications of Lyme."

I suddenly felt a burst of heat in my face. It was rare that I'd meet anyone who'd understand Lyme, much less someone who had experienced the loss of life that could come with it, the outcome that people seemed to be only slowly realizing was possible. And of all people, this guy. I avoided his eyes so I wouldn't cry and quickly handed him my credit card. "I'm really sorry to hear that," I kept saying quietly as he ran my card, but he seemed speechless. "Well, have a good night, okay?" He didn't seem to hear me, his eyes still glassy and dazed, staring at his monitor.

When I got to the track, the train was packed. Just before I took my seat, there was a loud boom, an explosion of sorts, and throughout my car the sounds of human panic rippled from audible gasp to scream. Everyone's minds were momentarily in sync: bombs. We were all thinking of the Paris attacks, I assumed, and how NYC could be next. Manhattan at that time reminded me of the days after 9/11—worse, even. The first few mornings after the attacks, when I'd go out to walk my dog I'd see more police than civilians on my usually sleepy brownstone-lined street in Harlem.

Another explosion, and then another. More sounds of horror from the train, louder this time. My heart went into a familiar racing, as I scanned the frenzied passengers in my car, all of their eyes looking a bit animal. But before we could be overcome by our fear, someone figured out the source of the booming. "It's just fireworks!" a gruff male voice in the aisle muttered. "The parade, people!"

A few of us looked confused, and another voice explained, "They have this Christmas parade in Poughkeepsie."

I heard another few voices, "Oh yeah."

And one more, "It's a really nice parade, you know."

That was my last public transit ride before the car accident. Two days later, I decided to take my car out, a 1988 Subaru station wagon that I'd bought at the end of the last semester, which had taken me to work several times a week for months—not to mention two cross-country trips in the summer. I was happy to be back in the car after so much time away, but the truth was I didn't want to deal with Alan and his relapse, or Lou and his grief and the reminder of the direness of my condition, or even the tension on the train of a sound, any sound. So much already felt unbearable that season.

I had no idea I was about to hit new limits of unbearable.

---

After spending the extra hours with my students that Friday, I drove home that evening with a particular cautiousness I had acquired since this latest relapse had begun. For the first time in two decades of driving, I was suddenly someone who strictly observed the speed limit. It was past seven and it had been many hours since the sky had turned dark, and only as my last student left did I realize that I had never driven home this late. It only gave me the slightest pause, though, as I was someone who drove cross-country and had done several legs solo, I reminded myself.

Besides, I had promised my students I would be there. *Extra office hours*, I had said all week, *I want to hear you all out. It's been a miserable semester for all of us and I am here for you. You don't have to come, but if you want I am here.* Two-thirds of the class came to see me that day.

I remember the drive, like one often remembers the moments before something monumental, in crystal-clear vision that feels indisputable. The cold brisk night air coming through a small opening of the window, meant to counter the thick blast of heat from the car's heating vent. The barely black of early evening sky, the many stars that were out that night. The unease of a snowless December, like slack tide, the taut serenity when you know something is coming. The mild murmur of my car's radio playing familiar oldies on an a.m. radio station I'd taken a recent liking to. The emptiness of the highway, a surprise for me, until I realized this is what Friday night looked like in the Catskills.

After an hour or so on the road, I was feeling a bit bored, so I left a phone message for Mason, my old graduate assistant at another university I had adjuncted at, who'd also been my recent cross-country road trip partner. He'd often check in with me, worried about my health. I remember rambling on his voice mail, *Oh hi, it's me, Porochista, how are you, hey I'm back to driving, long day at Bard but good one, I think I'm feeling better, things are getting good, want to catch up on your week, call me back, okay, bye, kiddo!* It was one of the rare times he did not pick up on first ring.

Mason had gone through one of my "incidents" with me already. He had been the first person I'd called when the Lyme relapse first hit me in November, when I'd pulled over on the side of the highway one rainy evening, suddenly feeling like I couldn't tell where I was after a long day of teaching. He was the only person I could think to call since he was always checking to make sure I was okay. He had met up with me at the hospital where they'd checked for a stroke with a CT scan but found nothing—*Probably like you say, it's Lyme*, the neurologist said lukewarmly. I didn't think twice about it. Hospital visits were to be expected for the Lyme-struck, after all. And that was when I'd simply called my doctor and he'd ramped up my supplements and suggested the cane.

I left another message that night for my friend Bobby, who lived only blocks from me, who was the gay Iranian American brother I never had but always wanted, and who had been so concerned about my health that season: *Hey Bobs, I made it through a really long day up at Bard and I'm driving home again! Yes, driving with Cosmo, all good! You know what that means—I think I'm feeling better! Anyway call me back!* It was also one of the rare times when Bobby did not call me back just seconds after screening his landline calls.

I put the phone down and glanced in the rearview mirror at my standard poodle, Cosmo, deep in sleep in the backseat. Just moments later, a giant truck burst from the darkness and completely overtook my lane, like a monster that absorbs you, full speed and confident, no hesitation in sight.

I felt two impacts and it took both of them to realize what was happening. I'd been in two car accidents before, so I knew well the sensation of watching one's self melt into a slow motion movie montage: here I am honking my horn, here I am praying out loud, here I am screaming, here I am accelerating and braking and nothing feels right, here I am spinning, and here I am stopped. And here I even am alive, it seems.

On the side of the road, the car smelled like it was burning and I turned to Cosmo, who seemed shook, but okay. We were okay, we had to be okay, which I thought would be how I'd say it all, but by the time 911 answered I was screaming and I'm not even sure it was words that were coming out.

It took two 911 calls. And a lot of waiting on the New York State Thruway. We were on the side of the road but on something that wasn't quite a shoulder. There were no lights. After some time I turned on my hazards and looked into the rearview mirror and watched more cars speed by, each seeming faster than the one before. Everything seemed black and gold, confusing, elaborate, deadly, and strangely a little bit beautiful. It was then that I realized this could be it—the odds of being struck again by a speeding vehicle seemed higher than us just being comfortably lodged there until help came. I calmly thought that this was the end. And it took me a second to fight the thought and dial 911 again: *Please. There are no lights, no shoulder, cars are speeding, we are going to die. Please.* I remember my voice was too calm for those words.

The ambulance finally arrived. It seemed like I had no visible injuries and perhaps only a concussion, but I refused to go to the hospital because I was informed I could not take my dog.

"What am I supposed to do then?" I asked, pointing to a blinking and panting Cosmo, who looked only moderately flustered.

"You have to abandon the dog, ma'am," the paramedic kept saying.

"Of course I can't," I kept replying even as they told me stories of many accidents where the pets were just let loose in the woods on the other side of the highway, as if to console me.

There was no way. I finally made them a deal—that I'd get checked out when I got home. "I promise."

The police officer looked at me like I was making a bad choice, and his gaze paused at the cane by my front seat, as if I already had a prop of injury perfectly on hand. "It's for my Lyme disease," I said, and he nodded blankly.

They called me a tow truck that took me all the way home—me with a shaking Cosmo in my arms, my head suddenly pulsing, the tow truck driver taking an interest in two things: my not being married and my name. Eventually we got into it: Iran and Muslims and 9/11 and the Paris attacks, and after I realized this man wasn't going to hit on me, I was so focused on not letting him hurl anything racist at me that I barely remembered the accident.

At one point he said, "I'm gonna be honest with you, you Arabs have not been my favorites, you know?"

I didn't correct him and just focused on Cosmo's fast breaths, which seemed synced with mine.

"You are all right, you know," he said at another point, which I tried to imagine was meant about my health but he of course meant me as a Middle Eastern person. "You're a good rep of your people, we need more of that. Especially after what just happened in France, you know."

I knew. But I took this as him taking a liking to me, which felt at least safer than where my mind originally went, and so I was grateful when he dropped me off in front of my building as a reward. They weren't supposed to do that, he told me. Then he rambled on a bit about how he had once towed an Iranian restaurant owner—"Iran or Iraq, one of those"—who'd then fed him for free. "Sometimes you guys are good" was his moral, apparently.

I nodded numbly, thanked him. By the time I got into my apartment, I had many worried calls on my cell phone.

*How did everyone know this happened?* I texted back my friend Alex.

He reminded me that I had made a Facebook post about it, when I was waiting for the ambulance—just as I'd apparently called back Mason, who said I sounded incomprehensible. I remembered neither.

Alex kept asking if I felt okay, and I did. I told him I survived, and I instead lingered on the tow truck driver and the season of xenophobia and all its perils, what was most on my mind.

*Please check in with me*, Alex kept writing.

*K*, I kept typing back. *K.*

I began to get sleepy.

That night I fell into the deepest, thickest sleep of maybe my entire life, but definitely since my Lyme had begun to relapse that fall. In Lyme relapse, I never get proper sleep; it always feels like that light buzzy rest of past drunken nights. But this night was different. I slept twelve hours. When I finally woke up I was on my couch, and everyone was calling and texting at once, wondering why I wasn't at the hospital.

*Why would I do that? I'm fine. I slept well.*

But people were reminding me that I wasn't supposed to have fallen asleep. That people are supposed to wake you up every few hours after a potential head injury, in case of a concussion.

*But so what, I'm fine now. Who says I have a concussion?*

But no one thought I was fine.

*Okay I'm not fine—I've been having a Lyme relapse.*

Soon my old editor was ringing my doorbell and at my side, taking me to the hospital.

Upon seeing that dear friend, a deep panic took over me. I had a feeling I didn't want to get into what came next; it was something I knew well and had fought to get away from. "I don't want to do this again," I cried into his shoulder in the cab. "Not another hospital. You know how many hospitals I've had in my past. Not this again, please."

My editor knew my story and reminded me this was different, that I had to go, that it wouldn't be like those other times.

At the hospital on the Upper East Side, I fell back to my old element, almost finding the waiting comforting, all the systems ones I knew well. I was explaining to my editor how it would all go, when he interrupted me. "But we're not here for the Lyme, we're here for your accident. Remember?"

It was sometimes hard for me to comprehend there could be room for anything more.

I had been to the hospital so many times for my Lyme disease, not just explaining but overexplaining, as if I had something to hide. Lyme is a disease that many in the medical profession, unless they specialize in it, find too controversial, too full of unknowns, to fully buy it as legitimate. It's thought of as the disease of hypochondriacs and alarmists and rich people who have the money and time to go chasing obscure diagnoses. For years I'd become used to dealing with all sorts of skeptics whether in person or online, but it never stopped being frustrating. I'd always catch myself getting preemptively ready to argue, feeling a defensive heatedness from years of impossible experiences with so-called medical professionals. It was always a risk, me getting into it, and yet I had no choice. I had the script down as if it was a script and not my reality. *My case is a CDC-level Lyme case*, I'd learned to say, which was true, hoping doctors would understand I was one of the small percent of Lyme sufferers who actually had the luxury of CDC recognition, what the rest hope for. *I'm not like those other ones* . . . I'd try to knowingly add, to speak their suspicious dismissive language, to let them know I was real.

After hours of waiting, my editor had to leave to catch the last train upstate. When the young internist finally admitted me, he was surprised by my cane, but I didn't want to get into a conversation about Lyme right away, especially without an advocate at my side. The internist examined me and wanted to give me x-rays or a CT scan, but I insisted on an MRI.

"I've had too much radiation in my life," I said. My Lyme doctor always reminded me to say this; he did not like me going through airport full-body scanning machines, either, to this day only getting pat-downs. I had just had a CT scan in November and I remembered even then feeling like I shouldn't be going through with it.

The internist asked, "Why?"

Here it went: the great downhill. "Lyme."

And there it came: his half smile.

And here it followed: my rage.

He ended up prescribing me some Tylenol and said it might be a concussion but that they could not do an MRI and that I was likely fine, and to follow up with my Lyme doctor.

I watched him walk away, and as I put on my shoes and coat, I saw him and a nurse laughing.

As I walked out of the ICU, I felt that old state of mind consuming me, taking me back to my time in so many other hospitals, and the anger at being misunderstood boiled up in me again, that feeling of not being taken seriously by those who had your life in their hands. All the many times, the people who shook their heads at Lyme, who looked at me with pity for my circumstances, who could barely stifle their rolled eyes. I'd tried to avoid this hostile world of hospital rooms and doctors' offices for years, but it haunted me. Here I was again, with something unrelated to Lyme and only two choices—to come clean or to hide, but I knew every decision would have something to do with my diagnosis. It amazed me that even after all these years, with all the time that had passed, as I managed to stay out of the medical system for the most part, and my fluency in their language still being proficient, that I could still be in this position—helpless, crazy-seeming, confusing, inconvenient, out of their norm, a problem. And not one worth the time to be solved.

The internist and nurse were still smirking as I walked by. I couldn't keep quiet. Long ago I had promised myself to keep my self-esteem intact in a medical system that had too often threatened to destroy it. "I hope you know I can see you laughing there," I called into the room.

"No, come on" was the call I got back.

"How can you prove that anyway" was another call I got back.

"I can file a complaint," I shouted, and just like that I was escorted by a nurse to a desk where supposedly people filed complaints.

For hours I waited, though I could hear the voices of family and friends telling me I needed to rest. Instead I watched every interaction with the people in the ER waiting area, which was remarkably vacant. Eventually they told me whoever I was supposed to talk to was not there till the next day, and so I left.

Like I was never there. The old feeling: like a ghost I walked out into a still Manhattan, taking a cab home with little idea of what had happened those hours.

That Monday my friend Bobby came with me to my last day of Sarah Lawrence classes—me showing up to class was already against the wishes of my department chair, but I felt with a friend's assistance I could get through that final bit of semester. I was in denial about what indeed turned out to be a fairly severe concussion; all I wanted was to hold on to some sort of normalcy.

*I'm fine*, I kept saying. *I have Lyme. I've had Lyme.*

But what I meant was I did not have room for this new thing that happened to me, this thing that was making the old thing worse.

I had been hit by a truck; I had hit my head; I was far from okay. But in too many ways, I could not afford this reality.

---

In the next few weeks of my life, I saw everything turn upside down and my life became more bizarre than ever. I could not tell where I was at times—at a stoplight down the street from my house, a neighbor said I asked him if I was in New York City and he walked me back home. I began hearing things—on the phone with my friend Laura, I heard a brisk swirling sound that I knew as the ocean in high tide: *Do you hear that, Laura? Why are we hearing the ocean now? What is this?* Another time, walking with my editor down the street, I suddenly was overwhelmed with a metallic flapping in my ear. *What the hell is that?* I screamed. All I could describe it as was the feeling that the illustration on the paperback cover of my second novel, a winged man, had lodged himself in my inner ear cavity and this was the sound of his giant metalwork wings that were scraping against my eardrum.

With a friend at my first NYU neurology appointment, I fell into violent convulsive crying when shown a chart like the ones you get during eye exams—not only could I not make out the letters, but they were flying at me in all sorts of directions, the entire room wobbling along with it. And my one solace, the computer that was for writing and communicating with friends and learning about the world, suddenly made me nauseated at best and put me in panic attacks at worst, as if its thinly vibrating frequency was suddenly amplified so that I could see its inner workings, its multitudinous networks busily announcing their duties, as if I could suddenly see the blood vessels underneath my skin. I was only allowed brief spurts of computer use—five minutes at a time for severe concussions like mine, followed by hours of rest, eyes closed, in the darkness. If I tried to do more, my body would punish me. All my senses had gone hostile.

What I didn't know for sure—though my doctor certainly did—was that I was having a major Lyme relapse now, with the accident and concussion nearly finishing me off.

---

By the time New Year's Eve came I was still in a haze and could not really register that this could be seen as a fresh start, that the holidays were over at least, and life in 2016 could now take hold. I wasn't mentally together enough to invite a new beginning.

Instead, I was focused on the logistics of my limitations.

I knew, for instance, that I could not go outside for Cosmo's last walk at midnight, when I usually did it, as there would be too much going on. I had never been in my neighborhood for New Year's—always leaving town in previous years—so I didn't know for sure, but I could imagine it.

But as usual, as the hours went by, I grew more and more lazy about that final little stroll. Cosmo's walks had been cut down by many times at this point, and he himself—I was still a week or so away from realizing he had suffered pulmonary contusions from the accident—had adjusted to his limited time outside. It was a miracle to me that I was even maintaining a basic walking schedule with him.

I finally got up, got him leashed, and got us out. I paused at our doorway.

I didn't know what time it was, but it was possible it was that wrong time.

Just as we walked out the lobby I heard my answer in the form of a husky laugh: *Five minutes to midnight!* went a tenant, a short shiny dress of a woman slung over a suited man.

How had I picked that time to go out, of all times?

But it was too late.

There we were in my Mount Morris neighborhood of Harlem, at the gates of Marcus Garvey Park, taking a longer walk than I had tried of late. It felt good outside to me somehow, far quieter than I'd imagined, the perfect temperature, the outdoor air somehow emitting *special occasion*.

I had lost my watch at the ER, so I couldn't tell on my own when it was coming, but moment by moment, out of sync, I began to hear countdowns. In the grand windows of my neighborhood's countless brownstones, I could see all the signs of celebration: disco lights flashing, groups of friends gathered around a table, champagne flutes raised.

3–2–1

*Happy New Year!*

Then all the sounds of joy. And pops. Gunshots maybe.

And there we were, woman and dog, alone, survivors I hoped.

The full weight of my aloneness hit me in that moment.

If I was going to survive it at all, this time, it was all by myself. And here, in this place that I had to reconcile with as home. If you face yourself properly, you also have to at some point face where you take up space.

I hadn't thought that way before. Far from it.

# 1

## IRAN AND LOS ANGELES

The one thing I do know: I have been sick my whole life. I don't remember a time when I wasn't in some sort of physical pain or mental pain, but usually both.

I was born in Tehran in 1978, infant of the Islamic Revolution and toddler of the Iran-Iraq War. Like many Iranians of their educated, progressive, Western-friendly upper class, my parents did not last there. My first memories are of pure anxiety, buses and trains and planes with my two parents, who I was cognizant were just two clueless beings—a twenty-six-year-old, a thirty-three-year-old, both often in a panic, occasionally in tears. Furiously I told stories to distract them, books the only toys we could fit into our two suitcases. Whenever I could, I took pen to paper and drew images and had my father dictate my narrative. It was not much, but it was something; storytelling from my early childhood was a way to survive things. Meanwhile I tried to ignore everything in the air—hot air balloons, helicopters, later fireworks and light displays—that resembled air raids and bomb sirens. I knew to hide my trauma, at least until we found a home.

*This isn't your home*, my father would say when we got to the US. *We'll be going home again one day soon.*

He said that for three decades. Now in the fourth, he likely still says it.

Years later several doctors, and also my own parents after they'd read some study, told me that if a child is exposed to significant trauma in the first three years of their life, they could have significant psychological repercussions later in life. PTSD. The brain of a child develops at high speeds, those first few years a time of rapid development. And as their brain develops, so do their emotions.

No one who knows this study has ever let me forget this fact.

Like most Iranians we ended up in Tehrangeles—almost. The portmanteau referred to the enclave in the West Side of Los Angeles that had become home to an Iranian diaspora, mostly refugees on political asylum like us, but also many other wealthy Iranians who simply felt at home in the Southern California of luxury and hedonism, far before a siege on their homeland would force them out. But while we'd spend weekends there at various kebab houses, driving around Rodeo Drive and gaping at rich Iranians flaunting their nose jobs and carrying designer shopping bags, and going on long rides to see their McMansions, we never lived there. We were forty-five minutes southeast, inland in the East Side suburbs, from Alhambra and Monty Park to eventually South Pasadena. My parents were no longer crying at least.

And so my father began looking for work, a process that has never quite ended. His accent canceled out his MIT PhD—and this could only mean adjunct professorships with little hope of much more (my father, well into his seventies now, is still adjuncting). At first we were a quick drive from Cal State LA, his first place of employment. It was in the library of that very college that he learned from a librarian of a city called South Pasadena, a place that was expensive if you wanted to buy but affordable if you wanted to rent, and it had an excellent public school system. That was all that mattered to my parents, my future still a thing not ruined in their eyes.

We lived in that greater Los Angeles suburbia from when I was age five through high school, the four of us, two upper-class fallen aristocrat Royalist Iranians, who left everything behind and raised two lower-middle-class children in a tiny apartment in South Pasadena, California. Even during our greatest financial struggles, my parents were too prideful to accept government assistance. My whole family shared a single bathroom and my brother and I shared a bedroom until I was seventeen. While my parents mourned their lack of wealth, all we cared about was the brand of normal that sitcoms, films, and songs of the eighties promised: a sort of oblivious California happy that could cancel out the news and its consistent airing of grievances against our homeland. Iran was never far from the media's lips, it seemed, and that poured into my home and playground life with equal heaviness.

I love that sad, simple apartment, but it wasn't until I was much older that I recalled the mold in the bathrooms. The cracks in the ceiling. The peeling paint everywhere. The suitcases in the closet always packed for an emergency, always packed for fleeing, presumably back to Iran. The way we'd have to negotiate lunch money. The times of true poverty that my parents lived in that they thought we couldn't see.

Decades later Los Angeles would be the only place I would contemplate suicide—not once but twice—the least healthy place in the world for me. In a way, I always felt it. My entire childhood I was sick with one thing or another, always feeling broken in my body. Depression was a constant. It would be many years until I realized that not only did Los Angeles have particularly horrible pollution but that Pasadena was one of the worst—a big bowl of dirty air thanks to car culture plus the mountains, a sponge absorbing all the smog. Good schools, yes, but we had ended up in the worst place for health imaginable—though for my parents, both relatively able-bodied, health never seemed to be much of a factor. No one paused and thought my preferred dinner of two Wienerschnitzel hot dogs or McDonald's chicken nuggets or KFC buckets was bad. No one stopped me from soda. Nobody asked what I was spending lunch money on, when it became clear I'd be bullied if I put Persian stews or cold kebabs and rice anywhere except in the trash. No one bothered to knock on my door those weekends when I stayed indoors every second, reading and reading and more reading, and writing furiously, still deep in the dream of stories, fully invested in being a "girl author" one day. I had notebooks devoted to plotlines, to vocabulary, to illustrations of my stories. No one bothered to say, *That's nice, kid, but you also need to go outside*. While I attempted community center classes in ballet, jazz, tap, gymnastics even, no one told me to stick with it and not to quit, when as usual I realized I did not fit in. The only place I fit in was not quite home but what was within our home: it was the desk, the chair, the pen and paper. It was the only place I felt well.

As a child I had severe ear infections and one that led to near deafness, though I never let on that I knew my hearing was dropping out, and I'd merely guess at what people would say to me. Occasionally I was right. Eventually I had surgery to put a tube in my ear. It was then that I had what I consider my first drug experience—what I used to joke about, but today I do indeed think of as part of early wiring that led me to drugs later in life, or a fascination at least with altered states.

The doctors apparently could not give me enough general anesthesia to put me out during the routine surgery, which resulted in my very lucid hallucinations. I have two memories of the whole ordeal: one is waking up, with surgeons in their green uniforms hovering over me, their movements and sounds ultraslow while behind them in the hospital everything moved in hyperspeed. This was only one of the couple times I woke up midsurgery, when they could not put me out—but the time I remember. The other memory is of afterward, waking up clutching a stuffed animal I didn't recognize in the waiting room, on my mother's lap, being told by the doctor that he had never met such a brave child.

"We couldn't put you to sleep—you didn't want to go to sleep, brave girl?" he said. "I've never seen such a brave girl."

I remember not understanding, feeling suspicious of the doctor. "What happened to me? What went on in there?"

"We fixed your ear, brave girl," the doctor said. He clearly could not say my name, something I was already used to by that point.

"I woke up," I said. "I woke up and saw."

Nods and chuckles, doctor and mother.

Later in the car I tried to convey to my mother that time had been altered in there, that it had sped up in parts, slowed down in others, and I saw it happen.

"It was the gas," my mother said. "They gave you a gas to sleep."

"I was awake," I said. "I woke up."

"Yes," my mother said wearily, "it was a drug. The drug made you see that way."

In just a few years, when Nancy Reagan and the *Just Say No to Drugs* slogan was as ubiquitous as *Merry Christmas*, my mind darted back to this memory—my mother telling me they had put me on drugs, that I had drugs and that they had made me see that way.

The end result of that: I became suspicious of doctors. And I became fascinated by drugs.

I had my own ideas. I tried to steer clear of our family pediatrician who always seemed a bit *off* to me. I decided the life of the body would be a secret life and that I was in it for the brain anyway. So many books to write, so many to read, so many words to learn. I was determined not to get tested for ESL again, determined to be one of the honors kids, determined to do justice to Mrs. O'Connell, my second-grade teacher who saw my handmade books that we made from the refugee days and onward and said, *I'll bet you will be an author one day.* And it was because of her that I agonized over submissions to *Highlights for Kids*—always rejected—and that I made my father buy all sorts of guidebooks on how to publish kids, all fruitlessly. But I had a work life and there was nothing that was going to get in the way of that, not even the flus and colds I would always get. And not even the tremors.

I never told anyone about the tremors, but that was what I called them, named after that word I had learned that referred to the earthquakes of my new habitat. I was still a few years from experiencing earthquake tremors but always greatly feared them. The first earthquakes I experienced were the ones in the body instead. They always came late at night, when my parents were supposed to be sleeping in the room across a tiny hall from my brother and me, bridged only by the restroom. The tremors always came after I'd listened to my parents for a while, their whispered conversation turning into a whispered sort of yelling and sometimes bangs, what I recognized as the sound of a rattling headboard.

My brother somehow always slept through this. I was always

deeply worried and deeply afraid to do anything about it—I could not make it out of my bed and risk being seen as witness to their fights. And my body, as if in cooperation with that notion, didn't let me. Generally a half hour into it, my entire body would go into a shaking, starting with my legs, a sort of violent involuntary rattling that would stop on its own after a while. Sometimes it felt like hours would go by. Suddenly it would be time to go to school and I had no explanation for why I felt like death. My mother never suspected I was not sleeping, never suspected I was listening to them, never suspected the tremors. Only once, one summer evening, I remember her taking my hands in hers and staring at them in deep concentration.

"What are you doing?" I asked, alarmed.

"Your hands are shaking," she said. "Why are your hands shaking?"

I shrugged. I had not noticed.

"It's all the writing," she said. "You need to stop writing so much. It's cramping your hands. Take a break." Later she said that about tennis, piano, pottery, anything I tried to expand myself in, but her jab at writing hurt most because it was the first thing close to a purpose I had found in this new world of ours. I remember feeling furious with that suggestion but knowing better than to fight with her. I would keep writing. At that age I already knew they could only control me so much.

The tremors seemed to fade as I entered my preteens, though that was when I cared less about listening to my parents' late-night altercations. Walkmans existed and I had convinced my parents to get me a Sony Walkman for Christmas one year, which felt like the greatest invention on earth, especially for someone like me who was so dedicated to a private life. I could listen to things without them interfering. Radio could transport me into all sorts of possibilities, and I was allowed cassette tapes here and there, Madonna, Michael Jackson, Janet Jackson, and Cyndi Lauper. While my father argued they were all obscene, he did not go so far as to stop me.

By my preteen years I was obsessed with late-night radio, where I could learn about what cassettes to buy, and where I could listen to the show *Loveline* on one of my two favorite stations, KROQ. It was a show on which they'd have a music guest, which is what I pretended I was in it for, but the main point of the show, love and sex advice, was the real allure for me. KROQ must have been where I learned about sex, as it certainly did not come from my parents. So from 11:30 to 1:30 a.m., I could put on my headphones and not disturb my brother across the room, pretending, when my mother tucked my brother in, that I was also deep in sleep, while I was actually learning about STDs, 69ing, spitting and swallowing, orgies and ménage à trois. She never bothered to check the headphones that I wore upside down under the covers. Listening and laughing along with Poorman and Dr. Drew seemed infinitely preferable to listening to the terrifying bickering of my parents. Maybe by that point I had learned nothing too terrible would come of it, nothing worse than the other bad things that seemed to happen in our house anyway.

The next time I remember being very ill also involved entering in and out of consciousness, though not so artificially as when I was drugged for my ear surgery. I was thirteen and it was a Sunday, as the concept of school was looming large, and I had just taken a shower. The shower was a hot one—I've always been partial to very hot long showers, especially when given the mammoth ordeal of dealing with my always-tangled, always-frizzy, impossibly thick black hair. But this time, in the fog of the bathroom, as I dried off, I felt strange. I felt both lighter and heavier, like I was being pulled up and down at the same time.

My eyesight seemed altered—visuals seemed both brighter and darker, strange shadows jumping in and out, all electric in their tone, nearly metallic in shade. I felt like the life force was being vacuumed out of me, from every opening in my body. I wrapped the towel around me and wandered into the living room and called out to my mother.

"What's wrong?" she asked absently, engrossed in the television, on her usual love seat perch.

I remember saying simply, "I think I'm fading."

The sentence did not seem to mean much to her, and so I struggled alone to their bedroom, not mine for some reason. Moments—minutes?—later I was lying on their bed, and my mother was shaking me and speaking rapidly in a panicked high pitch I'd never heard. "What happened?" I asked, as the colors of the room rearranged themselves into normalcy, the sounds falling into their more proper places.

But she was still red in the face and frenzied. "You fainted!" she cried over and over. My father was now at our side and he seemed very upset too. But I was beyond the realm of upset or anxiety. I felt more peaceful than I had in ages. In the days to come, I felt special as a fainter, as if I was a character from another world. It felt like an event to have a condition, especially since I was still months away from getting my period, the affliction that it seemed everyone I knew got to complain about.

This called for me being dragged to our family pediatrician who said it was normal for my age, which disappointed me. But then he brought up the possibility of me carrying smelling salts—which I'd only known from old Victorian epics and period cinema that I had at that point become obsessed with—and I was overjoyed. I eventually got them, but I never got the opportunity to use them. I never again let myself fully pass out—instead, when it came to that intense fading, the light-headedness, after a hot shower and often at malls, I would sit and place my head between my legs as instructed, and I'd always somehow evade it.

I liked that there was danger involved with me, that I was someone people could lose, that I could flirt with some other realm, that I was intensely fragile yet ultimately indestructible. I felt like a crystal ballerina, a porcelain swan, but most of all like a ghost. The haunting metaphor felt actualized in some part of me: a part-ghost at least. I had access to some other world, but I could be in this one too—I told myself that narrative. The narrative I ignored was the one where I should have also realized that it was the first time I got to feel like a woman—and that perhaps ailment was a feature central to that experience, the lack of wholeness one definition of femaleness, or so they would have you think.

---

When in 2009, almost twenty years after that first fainting incident, Dr. E, an infectious disease specialist who treated my Lyme in Pennsylvania, tried to get my entire history of illness, I recalled all these instances. It interested him, the ear surgery, the tremors; the fainting. He asked me if there was any chance I could have been exposed to ticks, and after much struggling I could only come up with one memory: when I was around six or seven, just a couple of years into our new life in America, my father took my mother and me hiking.

I remembered it being the Angeles Crest Mountains, because I loved the name—back then I digested American words like they were frosted desserts, and anything with "angel" was delicious to me. We were hiking through a forest that seemed unlike much of Los Angeles, and my father said they reminded him of the mountains in Iran. (My father tended to say that about most mountains.) We got to an area with a sign, a wood sign that I still vividly remember as painted brown with a very friendly-seeming yellow font that announced LYME DISEASE • BEWARE OF TICKS and some other fine print under it that I did not read. I remembered it because of the word *Lyme*. I was new to English and I had become obsessed with spelling, and I turned to my father and said, "They spelled *lime* wrong, didn't they?" And he had stared at the sign long and hard and my mother and him muttered back and forth and then shrugged off their worry, as they usually did, like people who had seen it all long ago. Ticks were the least of their concerns.

I remembered the Farsi word they used: *kaneh*. Which referred to a pest of some sort and is a sort of insult you could use for someone, someone more than possessed by a bite, someone who has lost it. My father explained it was a disease one could get from bug bites and I remembered thinking bugs liked to bite me—every summer my legs showcased constellations of mosquito nibbles. And then, as children do, I forgot all about it and we hiked. It's one of my few memories of hiking with my family—the Khakpours were seldom the outdoorsy sort. And it was eighties Los Angeles, when hiking at some altitude was the only way to get away from the thick brown veil of smog that covered the city more often than not.

It was only years later when a man who loved me took me hiking in that same area that I again saw the sign—this time I was with diagnosis, sick and shaking—and I thought back to telling Dr. E the story, who saw it as a *highly probable* origin story. And I thought back to the event itself and the calm of my parents and my own calm, probably thinking how on earth could a bite from an insect do damage.

Of all the things that could do damage—revolution, war, poverty in this new land—why would anyone think of a *kaneh*?

# 2

## NEW YORK

Ever since I can remember, I dreamed of escaping. Escaping *what* was always the question, but my life had been one of escape since I was born—revolution and war sent us through Asia and Europe and eventually to America. We were in exile, my parents always reminded me, we had escaped. It was temporary. But escape was also something I longed for in eighties Southern California, which constantly felt foreign to me, a place of temporary settling but no home. Everything was tan in a way my brown skin could not compete with. Everything was blond in a way my bottle-blond mother could not re-create, gilt upon gold upon gilt. Everything was carefree and smiles, gloss and glitter, and money to no end. We, meanwhile, were poor and anxious and alone. When my brother was born in our neighboring city Arcadia, California, in 1983, I watched his pink squirming body stowed into a giant felt red heart—it was Valentine's Day—and even stuffed in all that makeshift American affection I thought he didn't have a chance. None of us did.

As the tremors continued, as my body somehow grew smaller rather than larger—my mother always quick to slap my hand when I reached for the leftover cake batter the way sitcom kids did, her ritual baking more American obligation than motherly delight—I also began feeling a need to escape the body. All my few friends got their periods before they were teenagers, but mine waited deep into my first teenage year, on the brink of fourteen, like an afterthought. Everything about my body felt wrong to me, especially as California went from the eighties to the nineties, and I knew escape would have to be a real revolution of presence.

My mind always went to literal distance, eyes on the globe landing without fail on New York. It's hard to know if all the movies of the era did it, *Fame* and its many knockoffs, *Annie* and all the stories of rags-to-riches miracles in Manhattan, told me New York was the motherland for misfit creatives to thrive, for foreigners with big dreams, for girl authors. But I think where it really came from was my aunt Simin, who was the only living role model I ever had. My mother's world, as it sought to merge with the average American woman's more and more, spoke to me less and less—I found myself cooling from her endless mall outings, Estée Lauder free gifts, diet everything, soap operas, and department store catalogues. Instead my eyes went to my father's sister.

Simin was not only the first model of a woman I could aspire to, but also my first example of an artist, a New Yorker, and ultimately a sick person. My father's three sisters were all supermodel-tall and bone-thin and prone to throwing on amazing mod dresses and thick black eyeliner with a messy casualness, like the heroines of old French cinema classics, all vaguely tragic and uncontainable and iconoclastic. I loved Simin the most, her skeletal frame always bearing the strangest clothes hung like modern art, grays and metallic and blacks and whites, thoroughly avant-garde. Her dyed long copper hair was witchy and perpetually bed-headed. Her face was all bones and lines and perfectly crowned the whole look, always a bit haunted but in this punk, high-art way that I did not have the words for then but could now describe as *New York*. She was an artist and spent half her time in New York and half her time in Paris, where her artist husband and grown son lived.

She loved me so much, and she was someone who always pored over my drawings and paintings and encouraged me to pursue visual arts, which was the one vocation competing against writing for me, but ultimately one that I did not have as much talent in. Still, she told my parents I should put my art on postcards and pass them out, always giving me confidence at the most unexpected moments. At my most awkward phase, early teens, she announced out of nowhere at a family gathering that I had the "perfect profile" and she'd like to draw it. (She never did.) Like me she seemed to hate our extended family gatherings, and more than once the two of us would stay in the car for hours while my family attended some function. It was there that she told me stories about New York City, how she lived in an apartment in a big building, with a view of Fifth Avenue—I remember her telling me how she saw Princess Di in some procession out her window. She had so many stories about the art galleries and boutiques. She once advised me that the way to deter crazy people on the street in New York was to act crazier yourself, advice I never forgot and still use on occasion. She told me one day we'd be in New York and do it all together, and I imagined her as my first roommate, I as her apprentice, going to art parties and gallery openings, sharing chic little meals together, never sleeping.

It never happened because on one of her many visits, this time when I was sixteen and was just two years from moving to NYC, she had a persistent cold and visited a doctor, and it suddenly turned out she had terminal brain cancer and had only months to live. Within weeks she ended up in a nursing home near my high school. I never once visited her, haunted so much by stories of my parents' visits, where she would spin paranoid tales and would be lost in hallucinations and nonsense plots, as her brain deteriorated at an alarming rate and eventually killed her. I refused to believe we'd lose her, but she was gone just like that.

For years, I was left with these memories of her, and I wondered how long she'd been ill. She was emaciated in a way I admired. She barely ever raised her voice above a whisper, which seemed elegant of her. There was always something a bit psychologically off with her, which I related to. Her fragility was almost a look for her. And all those years alone, away from her Parisian family—what did her life look like?

There began my fear of the woman artist alone in New York, who lives a wild and unconventional life only to succumb to the most standard and conventional institution of all: death. Premature fatal illness getting in the way of not just the accumulation of your years—aging never had much allure for me until later—but getting in the way of your artistic endeavors suddenly felt like a huge threat to me.

The greatest tragedy of the disappearance of my aunt, who had visited us in California—a place I knew she hated, always tagging along to the mall or beach with a sort of pained weariness—and never returned to her artist's life in New York and Paris, was that there was work of hers that would never be done. And my greatest apprenticeship, my old role model, was no longer there. I had no idea what image I was going to carve myself in, but I began praying I lived to achieve a few dreams at least, premature death and illness suddenly feeling like contenders in the many things from terrorist attacks to earthquakes to serial killers that could get me.

I never imagined a tick-borne illness on that list.

Plus, it would only be a matter of time until I'd set about destroying my body in other ways.

---

I felt more dedicated than ever to making New York my home, and indeed when it came time for colleges, all but the few obligatory California state schools were on the East Coast, and if not directly in New York like Columbia and NYU, then close like Brown. In the end, the school I chose—that chose me—was a surprising one: Sarah Lawrence College. I only knew a few things about it. One was that its pamphlet in the mail had announced iconoclasm was deep in its culture: *You Are Different. So Are We.* I also knew that it had a creative writing program, and that ensured I could be in those magical-sounding writing labs called "workshops." And finally, I knew that it was known for its lesbians—I remember one of my first female crushes, a beautiful Middle Eastern punk girl named Kara who was the star of our school's jazz band, once mentioned wanting to go to "Sarah Lesbian College." There was much of sexuality I felt I needed to at least have the option to explore outside of the supervision of my parents.

Westchester County, New York: Sarah Lawrence College in green suburban Bronxville. Even though the student body was made up of about one-third boys, it was still the typical rich girl's liberal arts college. I was the only scholarship kid I knew for a while and too many students felt okay to tell me their dads paid for me to go there. I had never even spent the night at a friend's house—slumber parties were not allowed back home, just like teen magazines, makeup, and dates of course—and suddenly I was going to be on my own twenty minutes outside the city I always dreamed of living in. Part of it was not knowing what to expect; part of it was that very actualizing of the fantasy, the New York girl I had fantasized about being.

In my first week there I marched to the college bookstore, knowing that you could charge all sorts of things to your parents' account and it would all come up as "bookstore," and I purchased a full carton of Marlboro Reds. As I child I'd watch old movies and pose with crayon cigarettes in front of the mirror, always looking forward to a day when I could become an actual smoker; my first crush was the Marlboro Man and I imagined sharing drags with him somewhere on a horse in Malboro Country. So here I was now, with my own cigarettes and no one to disapprove, as it seemed the whole campus smoked in 1996. I took them to my dorm room and spent hours practicing in front of a mirror, smoking a quarter of a pack a day to ensure I'd be addicted in no time. And I achieved just that—this would be an addiction that would plague me for over half my life. It would also be an affliction I brought onto myself that every doctor later in my life would note as a serious disadvantage in battling chronic illness, the body ultimately unforgiving of that vice, they warned.

I brought a wardrobe of all black and it only grew—it was the nineties and black on a scrawny girl like me meant instant fashion and art. I couldn't understand the Sarah Lawrence girls at first—too many were from the Midwest and South, regions that did not make sense to me at that point, and so I spent almost all my time in the city. I'd often miss the last 1:30 a.m.

train back to Bronxville just so I'd be forced to stay out all night in New York City.

And there I was at semisweet eighteen, the girl in the neo–Malcolm X glasses and black turtleneck writing in East Village twenty-four-hour cafés all night, who subsisted on coffee and cigarettes and bagels, who had friends who were downtown artists and poets and writers. Soon I knew where to drink without being carded, and soon I was the girl drunk for the first time in her life, on St. Mark's Place, vomiting on the sidewalk, not a single gutter punk or club kid blinking an eye. (*I'm soooooo happy right now*, I remember slurring into my friend's arms, who had tried to stop me until she realized my whole point was to get out-of-control drunk that first time I properly drank, trying every single mixed drink on the menu of the cheap café.) Suddenly I was the girl at every reading at St. Mark's Poetry Project and the Nuyorican Poets Café, dragging notebook papers full of poems I'd never read but wanted to baptize in good creative energy by carrying them in those spaces. Suddenly I was the girl who was going to clubs alone, who'd only return to campus to sleep away a few remaining daylight hours. Suddenly I was the girl with New York City boyfriends; suddenly I was the girl making out with girls casually as if it was nothing to me, at good old Sarah Lesbian College.

And suddenly I looked sick—looked like we all did, as heroin chic had taken ahold of the nineties and certainly our campus. To look just barely on the wrong side of life and the right side of death was a desirable thing, my friends seemed to agree. And that worked, because just as suddenly: drugs. Since that first hallucinatory surgery experience as a child, I had not had a drug experience. But at Sarah Lawrence drugs were all around me, in abundance. It was both a golden age and last gasp of counterculture; drugs were a part of life for nearly every young person I knew. It was hard to resist. The girl I'd imagined, the one who smoked cigarettes and wore all black, and went to poetry readings and puked on St. Mark's, was of course a girl who dabbled in drugs. I remember reading *Go Ask Alice* back home in LA a bit feverishly, all its warnings sounding mesmerizing, like all the *This Is Your Brain on Drugs* ads that had a secret allure to me. I always wanted that escape, and before I could even escape the body, I realized there were easier ways to escape the mind.

Dabbled I did. Pot doesn't count, everyone would always say, and so I went back and forth with that at first. It did little for me unless I mixed it with alcohol, always a bad idea but my kind of bad idea, the centerpiece of so many lost nights and wayward weekends. But then I had a boyfriend who confided in me that he'd begun taking coke, and if I wanted to, I could come to a "cocaine party." Hard drugs were not so inconceivable to me, as my first friends at Sarah Lawrence were all in "MacCrackHouse," what they nicknamed the dorm hall MacCracken, which was known for its junkies. The first drug I saw done in my life—before marijuana even—was heroin, when I wandered MacCrackHouse's halls freshman year. I was eighteen and was being invited to watch friends shoot up in their rooms. It was all vaguely glamorous—everyone in black patent leather, the soundtrack mod and industrial, people exhausted and beautiful, nodding in and out, luxurious and wasted. I dreamed of trying it, but I never did—one reason being that I saw the negative effects quite rapidly (and I ended up losing three friends to heroin from that time to a few years after graduation). But cocaine was something else to me, something that somehow seemed less deadly, but nonetheless sparkly and dangerous. I agreed to go to the party.

And that night—in a cramped dorm room filled with students I did not know and had never really even seen on our campus, sitting cross-legged on a dirty carpet, with action movie soundtracks blasting in the background, and a CD cover with a rolled dollar bill and white dust on it being passed around—was the start of a casual relationship I had with cocaine all the way to one last bump in 2015.

"Am I doing it wrong?" I whispered to my boyfriend over the Pulp CD, as I rolled and rerolled the twenty-dollar bill, hoping there was no residue on my face.

He squinted to inspect my face like it was a science project—people could be so serious on coke—and said, "No, you got it. Someone is a natural!"

I was. I enjoyed it more than I thought I would—it was like coffee but the high was very positive for me then, no anxiety at all in the mix, and it lasted for just the right amount of time, no long trips to worry about. It also did not seem to addict me in the way I feared, but my love of it also told me—just as cigarettes had proven—that I was indeed an addictive type. The drug found me over and over my next few years—once in the form of a present from a socialite who liked me and left an eightball in a cracked chocolate rabbit on Easter Sunday morning, often in the form of bumps from someone's back pocket, and then in daily endless supply from a friend who became one of the premier dealers on campus, and who in exchange for hiding her stashes in my room gave me as much access to it as I wanted. I dipped in and out of lines and bumps, often using them when big papers and early mornings were involved. It felt like just another part of me that I'd discovered in New York City—what New York City artist girls did if someone else was buying, especially.

In one of my earliest encounters with email I wrote to my California hometown best friend that I had tried cocaine.

Wow, you have changed, she wrote.

No actually I'm just more myself than ever, I wrote back.

And because this was the nineties, Ecstasy and mushrooms soon came in the mix, and marijuana became another staple as the thing you used to come down from all the other stuff. On any weekend night, either at Sarah Lawrence parties or at a downtown club in the city post-internship hours, I had all sorts of substances running through me. And not only that—I was well aware my substances involved other substances. You knew your E was cut with either heroin or amphetamines. Sometimes even your pot could be laced with PCP or just a "cocopuff" (pot sprinkled with cocaine). Once I even accidentally smoked crack thinking it was a cocopuff. None of these were big events, eventually, just things you did.

*God Bless the Nineties,* I remember scrawling on the whiteboard outside my dorm room in dry erase cursive—I would have tattooed it on my body, except I was still some years from being able to afford tattoos. But it was the right era for me, I always felt, I still sometimes feel. Everyone I knew was an iconoclast, a misfit, so different that we never considered we could all be the same, never thought that if enough people owned "alternative," wasn't it just mainstream? Never mind. The halls were always vibrating with nineties conscious hip-hop or druggy rock, like Gang Starr or Pulp, and we were always on something. Time was always running out, but in the best way, semesters just a hurdle to another break that I didn't want anyway—why go home when you could be in the middle of all of it, whether on campus or in the city? Who needed parents, stability, goals, a future? I was alive, in a moment, for once. My friends were free spirits, losers, anarchists, skaters, punks, taggers, club kids, strippers, professional junkies. I have very few memories of getting any work done, but I did remain diligent about my New York City journalism internships, as they were more than anything an excuse to have a purpose in New York City. I got to tell people, *I work in New York City*, even if it was for free.

There's a photo my parents took of my first day at Sarah Lawrence. I entered that first day in cutoff jean shorts, Pumas, a white T-shirt and baseball cap, styleless ponytailed long hair, makeup-free—the suburban uniform of any Southern California nineties kid. It took only a few hours there to realize I stood out, and in not a good way. By the end of my time there, in a photo from my senior year, I had a calculated mask of red lipstick and black eyeliner, hair in a studied frizzy shag, neck and fingers covered in costume jewelry, a frayed leopard coat on top of a black leotard and leather pants. I was at least a dozen pounds skinnier, my skin a bit gray; I gave off an air of dirtiness, in all meanings of the word. I had become something else, something that I would have once been frightened of, and that was the point.

Along the way, I just barely made it from tipping over completely into the dark side. We didn't have exams, grades, finals, any of that, but we had the equivalent and I was never sure how I passed some classes. There was one semester in my sophomore year when I hardly went to class at all, always in the city, school just an afterthought. While friends of mine dropped in and out of rehab and took leaves of absence, I was proud that nothing got out of control for me. I couldn't have afforded it if it had, after all. I treated my breaks at home as "drying out" and would assume my parents had no idea what I was up to, as I'd spend the entire summer in a dark curtained bedroom with my staple hoodie hood up, sunglasses on, with an impressive selection of vitamins, sleeping too many hours a day, sweating all sorts of things out. I'd work some shifts at the local Urban Outfitters in Pasadena, maybe intern at a magazine a day a week or so, but I was mostly glad I did not have my own car and did not know my way around Los Angeles. I didn't feel as if I was from there. I didn't want to know its troubles. It was now just a place for me to buy time, before the showtime of Sarah Lawrence and New York City.

There was one time my body announced itself and nearly escaped, a time when I came close to losing it all. It was my senior year. I had come home from *SPIN* magazine, my final internship, on a Friday night, and my central campus dorm hall was raucous with sounds of indie hip-hop, a Pharcyde and Dr. Octagon sort of night, which had to mean my friend Missy was up. She was a refugee of the old MacCrackHouse, and she always had the best pot, the only drug she claimed she did at that point. It had become my post-internship ritual to go to her dorm and smoke cigarettes with her in the hopes of getting a hit or two of her bong. She always shared, as she did that night, but this time the pot hit me badly. My other friend, also a former MacCrackHouse character, Ace, was in the room, and he and Missy were fine, but I was very much not. For a while we all assumed I was having some sort of panic attack—I had had my first few panic attacks those Sarah Lawrence years, often tied to drug use—and we tried to ride through it.

"You're cool with us, you're cool here, all is well," Missy said, swaying with the bong in hand above me, as I reclined on her floor, everything spinning. Missy was very rich, like all the Sarah Lawrence kids, but she worked on the side, which meant stripping in Yonkers, something I could never get out of my head every time I saw her.

"Please stop freaking out!" Ace was less consoling. "I can't deal with it."

It was bad drug etiquette to freak out. Back then nobody went to the ER, or called a doctor, or turned to help. You knew the risks—if you were going to do drugs, then it was all fair game. Even death. Back then we lost friends, so it was never unthinkable but rather built in to the experience, the risk maybe even part of the thrill. There was no going back. You had to do your best to live, but there were no guarantees.

I realized after a while that both Missy and Ace thought I was on something else. At some point I must have gotten sicker, much sicker, and they were more panicked, hovering over me, demanding to know what else was in my system.

I yanked myself up, suddenly lecturing them, "Fuck, nothing is in my system. I'm not a junkie like you guys! Cigarettes are in my system! That's it! I didn't even drink tonight!"

Somehow I stumbled out and made it to the doorway of my own room, and I realized my vision was a mess and I was seeing things. Orange cats were multiplying. Everywhere and fast.

Before I knew it, I was with Missy and Ace again, demanding they call 911.

Orange cats.

"Why? Just relax," Missy kept saying.

Orange cats. Orange cats. Orange cats.

"We can't be here if the paramedics come, Missy," Ace kept saying. Fucking junkies.

I demanded they help me, my heart beating too fast, saying I needed help and there were too many fucking orange cats.

"What do you want us to do, we'll do it, but what?" Missy was panicking.

"Dude, Porochista, we know how to speed up your heart not slow it down," Ace said at one point.

Like the druggies they were, they disappeared by the time I or they or someone called 911 and the paramedics showed up at my building, the very center of Sarah Lawrence's sole quad.

For some reason, I could not get out of Missy's chair—I told the paramedics that I was stuck, that my grip on the legs was not something that could be undone. I even laughed for a moment that someone had glued me—and the paramedics were not amused, just lifting me and the chair outside the dorm and into their vehicle on the lawn, past half the campus standing in their Friday evening best. The red lights of the ambulance gave the quad a sort of haunted emergency quality, like something very bad was wrong, but what?

Me.

I was what was wrong.

I was still seeing orange cats.

In the ambulance they gave me oxygen, the first time of many times I'd be given it, and they asked about drugs. I told them everything, and I begged them not to report it to my parents.

"You are an adult," one of the paramedics said.

I was an adult. He was right. I was twenty-one at that point. For years, without even noticing, I had been an adult.

At the ER, various hallucinations continued—this time I saw the male receptionist in blue sequins and a beehive wig, the sort of illustrious drag my favorite performers at the East Village's Lucky Cheng's would wear. I laughed and laughed, as I struggled to breathe and the various monitors beeped soberly.

I had lost all sense of time, but according to the nurses I was fine. I'd had something more than pot; it had to have been cut with something. But they were concerned about my other vitals throughout the night and kept me there for more bloodwork. Apparently I had disturbingly low blood pressure, an erratic heartbeat, and a slight fever. "Have you been sick for a while?" the main doctor kept asking, it felt like every hour, over and over.

"I told you, I don't know," I would say, because I didn't. How did I know if I was sick? I didn't have a life of thermometers and chicken soup and a mother and a good doctor. In New York I had a life of endless nights, sex worker friends, drinking too much and smoking too much, doing drugs whenever I could, keeping whatever hours I could build internships and classes around. I had a life of New York City. I barely ate, often some beans and rice from the cafeteria, a pie, too much soda, fries. In the city I'd hit up vending machines at work, and if my saved-up pennies allowed, a lemon cookie from a bakery at Grand Central, which had become a sort of reward for me. Reward for what, who knows, but I was living a sort of life that needed rewards, I had decided. I was also too skinny—an editor at the magazine had once said I was the skinniest human being she had ever seen, and I, remembering my dear aunt, had taken it as a compliment.

So I had no idea what normal was. I never felt good. I never felt not sick. I told the doctor that.

He wanted to know how much I smoked. Too much, I said. Had I been anemic before? How was my thyroid? What were my periods like? Had I been tested for STDs? Had I been tested for Lyme?

I shrugged it all off to get out of there, wanting the night to end, hoping for anything that would allow me to leave, but having no good answer for any of it. No one I knew went to doctors. No one I knew was healthy. No one expected it. If you were alive, then you weren't dead. That was it. It was just not in our culture to care.

The dean of studies came by some time around dawn.

"Look, I'm sorry, nothing like this will happen again," I blurted to him, embarrassed and shocked to see him there.

He looked resigned and disappointed in a gentle fatherly way, and for a moment I worried I was hallucinating him.

"Are you really here?" I gasped.

And there came his kind, booming voice, "Yes, Porochista, I am really here." We had never interacted much, so hearing my name pronounced correctly by an administrator astounded me.

"Please don't tell my parents, please. They won't understand, they don't know . . ."

And there it came again, like a taunt if it hadn't been so true. "Porochista, you are an adult."

I was an adult. An adult who had been dragged out of a dorm room on a chair, a chair she thought was a gluetrap, who had hallucinated orange cats and then a drag queen in a shiny blue dress, who had no health history, who had only junkies for friends, who had taken something that was laced with something and had nearly lost it.

Long after he was gone, when it was well into the bright Saturday morning, they discharged me and asked if someone was going to pick me up.

I remember laughing. "I don't have anyone. This is it, just me."

I remember walking like I was filled with lead to the campus, walking as if swimming in a pool of rubber cement. Everything was slow and impossible. I had ruined my life maybe. There was something wrong with me, many things wrong with me.

And those things were drugs, I thought. Why not.

There are ways drugs can coat all sorts of problems. You can think of drugs as pain relievers, and most of them are in some way or another. The body is asking for something, and drugs deliver something, but rarely that thing the body needs. In the end the needs of the body are unheard and another need opens to be filled. Drugs make holes so they can fill them for you later.

And so it made sense that my friend Ray, who had never done a drug in his life, who was the first of our friends to get a book deal—the week of our graduation actually, just months later—sat with me that night after the hospital as I cried and cried, over not my health or the incident, but over the fact that he was going through every one of my drawers and removing baggies and pillboxes and rollies and whatever else he could find to destroy. I cried out like a wild animal in pain when he flushed the grams of coke that I had forgotten about under my bed, flushed them down the toilet with a big smile on his face.

"Fuck you, that's several hundred dollars, you know," I bawled.

He knew of course that I never paid for anything. "You need to live. That's what I am here to help you with, Porochista."

Ray had already gone through one of my worst crises with me just the year before, when I'd emerged from spring break battered and broken. I had gone with a not-so-close friend and my boyfriend at the time to Martha's Vineyard, to my not-so-close friend's father's estate. We drank and did various drugs for days and days. In the midst of a rager one night, I was sexually assaulted in my bedroom by two men, while everyone else—my boyfriend included—partied in the main living room. When my boyfriend found us, the two men pretended it was a ménage-à-trois-type scenario, nevermind my ripped clothes and tears. We left the island at dawn and our relationship lasted only another week. The worst part of all might have been the rumors swirling around me when I got back to campus. Everyone in that tiny campus of one thousand knew something had happened to me, something big. I pretended not to notice the gossip, those eyes, the smiles. I was changed, I was tainted, I was scarred, and only Ray knew how to comfort me in all that chaos. I became severely depressed and plagued by nightmares of suicide, but Ray was always there to hold my hand and remind me that I could survive it, that I would have to survive it.

"You need to live," he said to me then and he said to me again, still several years from losing one of his best friends,

Rodney, to heroin. Rodney was only a pothead when I had my fling with him, which was already too much for Ray. "You need to live, okay?"

I knew that was true. Some of my friends who would later lose their lives to drugs were still alive, though just barely. I knew I had to live because, well, I didn't have the imagination to think otherwise.

---

That infectious disease doctor who would eventually treat me in Pennsylvania nine years after college, Dr. E, had wondered about my life and my history with ticks. When he asked me about my time at Sarah Lawrence, I looked at him blankly. I had never thought or even heard of Lyme disease in that way until that one evening at the ER my senior year, but I did know this: I spent a lot of time lying around in the grass. Lyme usually lurks in tall grass and wild meadows, so manicured lawns should have been fine. But often when I think of Lyme, I think of those first months of spring each year and how all my friends and I spent them as scantily clad as possible, often in bikinis, splayed out on the quad lawn, our reward for having endured winter, a taste of summer freedom to come. And I'd get bites, all sorts of bug bites, but what did bites mean in a time of hickeys, and even more so a time of my friends' track marks, when the only thing it seemed possible to die of was a drug overdose, with suicide a possible second? I did not think about bugs during that phase.

The infectious disease specialist told me that Westchester had always had high Lyme rates, but that he did not think this was where I contracted Lyme. He bet hiking in Los Angeles as a child was more likely than in college, even when I told him I was a mess in college, that all sorts of things went wrong, that all sorts of things could have gone much more wrong.

---

Only months out of college in 2000, I ended up heading back upstate periodically, in areas that I would later learn had more concentrated Lyme problems, such as Dutchess County and Stanfordville. It was there that my boyfriend Cameron's mother lived with Cameron's stepfather in a cross between a cottage and a castle. They were very wealthy—the stepfather was a retired surgeon, and Cameron's mother had suddenly come into money as a real estate whiz kid. Her first real job was a junior stint in real estate, where luck led to her selling a notorious golden cluster of Riverside Drive developments in her first years in the business. She had paid for Cameron's Ivy League education in cash, she'd say, not shy about it, not shy about anything really.

Going to the country to stay with Cameron's parents was a great escape from our life in a tiny studio in an East Village high-rise. Cameron's mother would take us apple picking, on picturesque hikes, to Culinary Institute restaurants. For a brief time we'd feel like New Yorkers who had made it so big we had earned leaving New York here and there. In reality, we were young journalists living large on someone else's dime.

But after a few of our visits, Cameron's mother developed Lyme disease—apparently she had gotten it from her dog, some kind of lapdog mutt, who had it too. She hadn't known what it was and had let it go, and she had rapidly developed neurological complications as a result. This was the first time I was properly confronted with someone who had contracted the disease, and it was hard to listen to her stories of the constant ups and downs and think that this had just come from a tick bite. We'd end up assisting her and she seemed mostly herself, but then suddenly she'd be off—staring into space, apologizing for scrambling words, throwing fits at her mind going blank, all things I would only come to understand a decade later.

"I'm never like not like this," she'd stammer with tears in her eyes. "I mean I'm never like this. I mean . . . both."

"You'll be fine," I kept telling her.

"Will she?" Cameron would ask me in the quiet of our guest room.

"Why didn't antibiotics work? Isn't that all you need with Lyme?" I'd ask him again and again.

Cameron would shrug, and sometimes in his passive indifference to his mother's state I wondered if he actually believed her. When we had initially started dating, he had told me, *My mother would do anything for attention*. It didn't seem impossible.

"Not a single doctor can help me," she'd mutter, which only made me more suspicious. How could doctors today not be able to help?

I'd try very hard to recall my coldness to her over a decade later, my inability to channel full empathy, my distance from whatever it was that was happening to her that I felt so far away from, so I could understand better when it all got turned around on me.

---

During that period, as we watched Cameron's mother get loopier and more pain-struck by the week, we didn't once consider that Lyme was something we also could get. She would beg us to check our clothes for ticks—*you don't want me to happen to you, trust me!*—and I ignored it, still having no real idea what a tick was, having no worry in the world around that issue.

Our longest visit was for a little over a week after 9/11, and it was a time when ticks were far from our minds. We killed the days outside, needing nature urgently, and never once did

I think that while my life had been altered by that one danger we'd narrowly missed, there was another very different danger that would strike instead.

*We're so lucky*, I whispered into where I thought Cameron's heart was as we slept in a tight embrace the night of 9/11, the only night we ever slept locked together that way. The feeling of being lucky never left me those days.

---

It's almost too poetic, the narrative that Lyme had come to me then—perhaps the writer in me believes it did because so many loose ends of events would be neatly tied there amid the tangles of character and plot. When I would later talk through my history with Dr. E, I would reflect on how I marked the passage of time in my youth: there was always someone, some man in my life. And this soon became the dividing metric of how I ranked my suspicions of when and where my Lyme disease came from. Did a part of me wish to braid the threads of story so neatly that way, or did I want someone to blame? Or was I more comfortable thinking someone else was there with me the whole time? Or was it the only organizing principle my mind would give, as the settings recycled themselves over and over, as my options grew more and more limited? Home for me was most likely in someone's arms, no matter how short-lived in the end; my location was so often defined by a relationship.

Case in point: my next serious boyfriend after Cameron took me deep into another neck of Lyme country, another affluent part of my world I'd never have access to otherwise. *In fact*, I'd tell the infectious disease specialist, *I've mostly blamed my Lyme on my Hamptons visits in the late springs of 2005 and 2006.* I wanted to blame it on that because I wanted to blame it on my relationship with Alexander, the only man I dated twice and the only man I never spoke to again, the ex who is now married to one of my oldest college friends, whom he met online months after we broke up. I don't have any good thoughts about Alexander.

He was a steel heir, and his grandparents were Hollywood aristocracy who owned a major estate on East Hampton. He lived in a tony part of Brooklyn and I had basically moved in with him in 2005. We'd spend long weekends, like much of young privileged New York of the mid-2000s, at the Hamptons. It was an intimidating performance for me, one I never got used to. We had this giant mansion to ourselves—it even had a movie theater that was almost as big as a commercial movie theater, and we'd drink bottles of expensive wine and fall asleep to Fellini films, just to pass the time. I had no idea then that the Hamptons were third to only Connecticut and Upstate New York for Lyme problems at the time. It was a period in which I wore bikinis and shorts and tiny dresses as much as I could, and we'd often lounge around in the acres and acres of yard the property had, just wasting hours like I imagined all rich people did with their time.

I felt bored by the relationship, but somehow also *chosen* by it. But maybe I was chosen by something else.

If I had contracted Lyme while living in Alexander's world, I felt it would finally give me something unforgivable to blame him for, reasonable or not. He didn't know that the woman he later married was a friend of mine, and they got together some time after we were broken up. But I'm still angry when I think of him and those long nights of champagne and starlight and candles, the hollow charade of it all.

He would ask constantly if I could see myself with him long term. *I want to hear you say you want to be with me*, he'd say over and over.

For some reason I would always hesitate. *Why do you need to hear me say that? I mean, I'm here. With you. We're together.*

He'd shake his head so knowingly, as if he knew better than I why I hesitated. *But I need to hear you say it, I need to know you know it. I'm thinking of a life with you.*

And I'd have chills considering that proposition, at twenty-six feeling so many lives away from that decision. And especially with a man like this.

*What's wrong with me? What is it that keeps you from me?* he'd rage at me sometimes, when I was like a distant star, blinking from light-years away.

*I don't know what it is, I don't know,* I'd finally say in tears, feeling terrible for not just him but us, and for the fact that I couldn't make a decision with my life that would obviously be good for me. At least on some level.

I began having panic attacks and insomnia—two symptoms that are now the hallmarks of a Lyme relapse for me, but at the time all I could assume was that they were due to Alexander and my inability to commit to him. They were from his hand always on mine, the way his eyes would try to read me, the never-spectacular sex, that damn estate and its endless luxuries, this world that young, beautiful, rich New Yorkers were entitled to—why on earth would I have been thinking about a tick in all that?

But when he broke up with me—something I did not see coming—I didn't feel relieved. I felt like the brown girl rejected by the white prince, an audition gone wrong, because I was just too terribly me.

Is it possible that I feel most burned by Alexander because my chance with that life of the chosen few also could have been the thing that could have saved me? After all, to this day what mainly separates me from healing is financial considerations. Could there have been another life that would have brought with it the opportunity of a more complete wellness? Is this why Alexander was the most worth resenting in the end?

It's the best theory I have. There are other things I observed, secrets of his family, but for now this theory is enough.

---

Most of the following years of my twenties were miserable, a seesaw of struggling to survive in New York and then running home to LA and then escaping back to New York. I was always very thin, out of anorexia or poverty or something else, but didn't have the mind-set to think about it. I would often come back to New York City with no intention of returning, remembering it as a place I thrive, knowing Los Angeles as a place I disintegrate.

In 2005, I told myself I was going to make New York work again. I had $1,300, $800 of which went to rent immediately—I rented out a tiny bedroom in my old Sarah Lawrence friend Julie's apartment in the East Village. The rest I stretched out. I ate only ramen, prunes, and graham crackers from the corner CVS. I kept growing thinner, my whole body sharp and birdlike in a way that's often celebrated in New York as fashionable and edgy. I began helping a friend launch a luxury lifestyle and high-end fashion magazine, which a sheik in Kuwait wanted to start up. This became my life, these long brainstorming meetings in my friend's Chelsea apartment—he also too skinny, both of us barely ever eating in front of each other. We were miserable.

During that period, late in the fall of 2005, I got in a car accident, a bad one. I was the passenger in a taxi in a multicar collision involving a driver on a cell phone and a traffic light out. I was banged up pretty badly, but the main injury was a part of my face was torn. At St. Vincent's I was transferred immediately to burn trauma and they cut through my paisley silk dress and brown fishnets—the nicest clothes I had, which I saved for parties—and sewed up my face and applied plastics and removed broken glass embedded all over my body with tweezers. For months, my life was physical therapy and even talk therapy. For weeks, my skin shed that glass, leaving behind little starlike scars.

I grew ever thinner, panic attacks came nearly daily, and I could barely sleep. I felt these symptoms were likely a product of the accident, though the whole time, I kept wondering if there was something else going on. But this was the one time I went to see any doctors in my twenties thanks to no-fault insurance, and all my providers focused on the accident. Xanax prescriptions were pushed on me, and soon I couldn't tell if I was depressed, addicted, messed up from the wreck, or something else.

I'm pretty sure it's just the accident, I remember writing a friend in an email. I mean, an accident on top of being very poor and not eating well and not having a proper job (the sheik has not paid us yet) and feeling anxious about being in my 20s in New York. The good news is Xanax is helping and they say the only downside is addiction. Which, I mean, I get addicted to things, fine, but it's been helping me sleep and even eat and sometimes helps with panic attacks—so what if I have to take it for the rest of my life? Do I really need to think about that right now? Haven't I been through enough hell?

My friend had written back: You just need to be careful not to add problems to problems, you know? If it's just the accident, time will heal it. Give it time. It will go away if it's just that.

I had written: Who knows what it is? I mean, I'm pretty sure it's that, but I don't know how to feel. I don't think I feel right, but when did I?

I don't remember her writing me back.

I didn't feel right, that I knew, but I also knew it had been years since I'd understood what that felt like.

# 3

## MARYLAND AND ILLINOIS

I made many legitimate attempts toward a life in New York, but the one thing I avoided all those years was serious writing—or at least any attempt toward the book I always imagined I had in me. Freelance journalism had been in my life since I was a teenager, and I had a bar column fresh out of college at *Paper* magazine, but nightlife writing and the occasional celebrity feature was no longer something I could tag *writing*. September eleventh was what pushed me to grad school, the Johns Hopkins Writing Seminars. I applied just a few weeks after watching the second plane hit the tower from my bedroom window downtown. In a matter of weeks Cameron and I had moved uptown to the Upper East Side on Eighty-Sixth between Second and Third, not the "nice part," mainly to get away from the view of the wounded skyline and also to live a life we couldn't recognize in the place of the ever-elusive optimistic fresh start.

All my unease in that period added up to me taking the GREs and applying to grad school. I didn't know what I was doing or why I was doing it, except that I was born to be a writer and there was no more time to waste, I thought. What 9/11 did was give us the gift of time running out; it put limits on things, stamped expiration dates on us. Suddenly I watched the news, the clock, the calendar. Age meant something. And I looked at my life just months before with a sort of mix of horror and envy—it had been a time of luxurious depression that would make me take aimless Manhattan walks all day, apply for a couple jobs, eat something, and then take Tylenol PM for the remaining eight hours Cameron was at work, rising out of that viscous slumber just as he'd arrive home to take me out to a different trendy dining spot every night. A consolation prize, an attempt at playing grown-ups. But after 9/11 I couldn't bear to live like that. At least grad school seemed like a proper way to make something of the many hours in my day.

By the time I got into the program, starting in autumn 2002, Cameron and I were over, to the shock of our circle. We had been a good pair, he a big and sturdy gentle giant, me a wiry and anxious and enthusiastic eccentric, both of us with an eye for fashion, both of us consumers of culture, both of us writers from birth. The breakup was hard, but he still moved me to Baltimore, where I was granted a year to make something of myself.

In Baltimore a certain old trouble—the one I had a hand in, addiction—followed me in a different form: it was where Xanax first came into my life, my first benzo, the beginning of a sort of end for me, and what I'd take back in New York. Baltimore I fell in love with, but the program was hard to love. There were only ten of us fiction writers and nearly everyone had gone to an Ivy League school. I was the art school girl. I was the only woman of color. I was the only one from New York City. The lone party girl, the girl everyone could tell had a checkered past. It took only a month for me to get involved with the other more troubled member of the class—a good writer, another writer of metafiction named Jim. He was an alcoholic from Connecticut, another Ivy League boy, brooding and beautiful and obsessively self-destructive. Long nights at the local bar made sense to me in the context of Jim, in the context of our program, in the context of Baltimore. He had no intention of making me his girlfriend and reminded me of that often, but we were a dedicated fling. *I can never tell if I'm attracted to you or not*, he said to me more than once, after some hours of empty sex. I found myself always wanting drugs, constantly asking around about them in a casual but persistent way, but the drug I got on was not the one I expected.

I started to sense something was very wrong after reporting into the program. I suddenly found myself mentally shambled in the mornings, almost feverish. I'd wake up with a million ideas racing all at once—ideas for stories, ideas on how to deal with Jim, ideas on my future, all sorts of conflicting and incoherent ideas on someone's workshop piece. It was like the opposite of a hangover—I felt as if I was short-circuiting not from burnout but from burning too brightly. I'd have to lie in bed for hours upon waking, smoking slowly, and just watching the millions of thoughts race by like an Olympic event. I'd blow smoke rings into the heavens until the mental rush would die down and I could get up. I began to dread mornings so much that I could barely go to sleep.

It was fashionable for the writing grad students to go to the counseling service, so I already had a therapist, but he became very worried about my state and suggested a psychiatrist. I obliged and met the next week with a very Johns Hopkinsian doctor, in that he looked like black-and-white newsprint, sober, gloomy, intense, scientific. He said I was "hypomanic" most likely and that bipolarity could be possible, but that most likely it was mania and it would pass. But it still required medication.

I had only taken one other psychiatric medication in my life and that was Wellbutrin, which in the late nineties and early 2000s was given mainly for smoking cessation. My friends at Sarah Lawrence all casually took it for some months and it prevented them from getting any of the negative side effects of smoking cessation, like weight gain. When I began dating Cameron, my smoking bothered him so much that he begged me to do whatever I could to quit, and so I tried Wellbutrin. I imagined it would do little other than magically shut off my appetite for smoking, but within a few hours of taking my first dose a dread unlike any I've experienced overwhelmed me. At dinner with Cameron that night, I squirmed and scratched myself and could not stop the feeling of wanting to jump out of my skin. I remember eying the knives at our table with too much interest and imagining carving an opening in my skin to crawl out of. When we got home I cried for hours and Cameron begged me to quit the Wellbutrin immediately, but I decided to do what my doctor had suggested: *Give it some time. Three weeks at least. You'll need to adjust.* It took a long time—I remember going to work and still going on smoking breaks and asking coworkers if they'd seen anything like me, as I shivered and shook my way through a cigarette, and they all just told me to give it time. I did and after a few months I quit it and quit smoking too. That was that.

So the introduction of a "mood stabilizer" seemed both novel and worrying. Still, Trileptal rather immediately did its thing: it slowed me down. The doctor also gave me a Xanax prescription for my high anxiety, to take as needed. Suddenly my signature maximalist, hyperkinetic prose turned minimal, choppy, tense, imitation Carver or Plath at the end of her life. My personality changed: I was less interested in sex. I was tidy. I was protective and quiet. I was under control.

And the drugs somehow allowed me to finish that year.

What I didn't realize was that I would not be returning to New York at the end of that year, but staying on an extra year—I had won a fellowship, given to four of us ten to continue. Jim had not gotten it, though he was clearly the best writer by far. But Jim's self-destruction was no secret to anyone, and especially not our professors, whose encouragements seemed to go nowhere where he was concerned. I was the other type: maybe not as naturally talented, but it was no secret that I had been hard at work at this for two decades.

When I started my second year in Baltimore in 2003, I felt haunted, nervous about the memories of the year before—from the destructive doomed relationship to the medication that stopped my racing thoughts. I was off the medication now, but what if everything turned bad again? It couldn't, I told myself, and I did my best to make the second year look as different from the year before as I could. I moved to the opposite side of town, from scruffy Charles Village to uppity Roland Park. I even took a part-time job as a hostess at a French bistro, and—because I didn't have any friends, either in the fellowship or out—I decided to go for a lifelong dream and adopt a dog. Not just any dog, but a retired rescue greyhound. I bought a copy of *Retired Racetrack Greyhounds for Dummies* and contacted the head of a local chapter. Soon I was attending one of their meet-and-greet picnics in Baltimore's nearby Guilford Park.

It was there that my first dog chose me. Kingsley was a white-and-brindle old guy, eight years old at that point, easily the sweetest of the bunch, and he immediately came over and cuddled up to me.

"He's chosen you!" Leanne, the chapter head, exclaimed. "Lucky boy, so lucky for a senior to find a forever home!"

I wanted to tell her to hold on a second when I realized he was actually bleeding into my lap.

"Oh no, his nails again," Leanne said. "He's got some work that needs to be done."

I wondered what work.

"Well, that and his limp," she said. "He's got Lyme disease."

I didn't blink much of an eye at that, though I hadn't known animals could get Lyme.

"It's nothing," Leanne assured me. "You just have to pop him some antibiotics whenever it acts up and that's it. We've just been waiting for him to get to a forever home so it can all be sorted out."

I had a moment of panic about not just taking an old dog, but a sick dog too.

"I would not call him sick in the least!" Leanne cried. And she proceeded to tell me a long story about how his father was one of the biggest track stars in Florida, a celebrity sire.

I agreed to take him home for a few days to test it out, to see if I could imagine a life with Kingsley, but I was a bit terrified. When Leanne dropped us off, I realized I had very little idea how to take care of a dog.

Nonetheless, over those days, we grew into each other. He went from being an awkward alien in my space, as if tiptoeing on eggshells lest he get too comfortable and thereby kicked out—that was the narrative I assigned to him, at least—and I went from someone who kept petting him to someone who tried to go on with her life, except for stealing glances at the lounging old boy who had suddenly made my couch his.

Kingsley's Lyme kept coming back, and, indeed, antibiotics were part of our life. I became a pro at hiding doxycycline pills in folds of peanut butter. At that point it never crossed my mind that this was a disease I could be in danger of, and I never imagined what it would be like to suffer as he did. Occasionally my mind would go to Cameron's mother and her case, but it felt so many lives ago. I accepted it as some minor part of my life with my dog, as I didn't have access to his grievances about it other than some basic motor impairment.

But Dr. E, the infectious disease specialist I would see many years later, had questioned me in great detail about my time in Baltimore. I just gave him a few answers.

Kingsley became an ideal writing companion for me, along with an occasional dose of Xanax as a sort of writing vitamin, and I managed to finish a draft of what would be my first novel. By the end of that year in Baltimore, I had written an investigative feature that appeared on the cover of the *Chicago Reader* and I was offered a job there as a staff writer in 2004. Chicago seemed like the ideal move, especially because I had no other plans.

I had also formed a noncommittal long-distance relationship with a man in Chicago by the name of Jerry—we'd met during my Northwestern journalism fellowship the summer before. So Kingsley and I, with the help of Jerry, made the trek to Chicago, for an apartment Jerry had found me. He was in Wicker Park and I'd be close, in the hip but still not quite up-and-coming Ukrainian Village.

Ashland and Superior turned out to be a mix of desolate and dangerous, and as much as Jerry assured me that many of his friends lived there, I never felt at home. Similarly, from the moment we moved there Kingsley seemed to get sicker and sicker. I'd go out to dinner and come back and the front room would be slick with his watery diarrhea (the first time I knew Jerry was not a keeper was when he panicked at the sight of it and left me to clean the mess all by myself). I had very little money to take Kingsley to vets, but when I did they could not determine what was wrong—either anxiety or Lyme.

I started to feel like Kingsley was seeing ghosts—in his

constant illness, I'd often catch him panting and staring at very particular corners of the living room, and then gasping and shutting his eyes, and repeating that cycle a few times until the next day at around the same time. It was when I made a crack about that to a new friend of mine that I was made aware of the fact that this space used to belong to junkies.

"You didn't know?" she said, laughing. "The whole neighborhood is full of drugs."

I told her it was hard for me to notice after New York and Baltimore.

"I mean, I thought you knew! Your building in particular was a sort of drug den. I heard a rumor—just a rumor—that someone died of an overdose there just before you got here. Or something else? Someone killed someone over drugs. Someone died, anyway, involving drugs."

My stomach sank. "And that's why it was so cheap."

My friend shrugged.

Soon enough Jerry and I broke up, and I started to wonder what the hell I was doing there. Most of my Chicago friends were from LiveJournal, where I'd mainly complain about my job, my lack of motivation to edit my novel, my dog's ailing health, and my discomfort in my potentially haunted apartment.

It was then that Sam and Cindy came into my life—a couple from around the corner, who'd been introduced to me through a friend of Jerry's. Sam was my age, a writer, too, but with a wild background: he was an ex-con, having spent most of his twenties in prison for heroin trafficking, and he now wanted to be a novelist and a fashion designer. Cindy was his newlywed wife, a Kate Moss lookalike who started modeling. They started offering to meet me for walks with my dog, and soon they'd come over, Sam full of conversation and Cindy, while at first uncomfortable and vacant and bored, eventually conversing too. They became my closest friends in Chicago.

*You are family with us!* Sam told me just days into knowing me, and he repeated it often. *Don't you dare forget your family!*

It was nice to have family out there. Sam and I had a lot in common. We were Middle Eastern for one thing—he was Palestinian Iraqi. We both loved punk rock and hip-hop. We both loved books and writing. Sam had this extreme confidence in all of us becoming successful—and soon. And I believed it.

He would always take me out to dinner with them and refer to Cindy and I as "his harem." *Bring my harem girls whatever they want*, he'd wink at the waitress, who no doubt did not expect that much charm from this short guy covered from fingers to chin in jail tattoos, nor the wads of cash that tumbled out of his wallet at bill time.

Eventually I knew better than to even offer at the end of a meal—he would always slap my hand a bit too hard. *When you publish that bestseller of yours!* he'd always say.

For months they were the people I saw daily, but I realized there was a lot I did not know about them. It took a drive to the vet—yet again for Kingsley's mysterious ailing health—for Cindy to confide in me. She had offered to pay the vet fee and I had refused.

"We have a lot more money than you," she kept saying.

I knew they did. I had been to their sprawling flat in an old factory building, and it was full of antiques and valuables, a weird Gothic museum of curiosities that made my apartment look like a teen boy's dorm room.

"Do you know where our money comes from?" she finally asked as we pulled up to my building.

I realized I had very little idea, though after New York I

just assumed every creative type was a trust funder. Still, they didn't seem it. And they couldn't be drug dealers, either, because Sam had said he would never risk jail again.

"I'm an escort," Cindy said plainly, as if ordering fries from a drive-through. "A prostitute," she inserted into our pregnant pause, as if she wasn't sure I knew what that meant.

I knew. "Does Sam . . . ?"

"Of course Sam's in on it," Cindy said. "It was sort of his idea. Sort of. He's even there."

"He's there?" I did not understand.

"He's in the room next door," she explained. "With his gun. He doesn't usually have to get involved, but he's there for me."

I nodded as if it was all entirely normal.

She popped open the lock on my side of the door and handed me Kingsley's leash. He was up and ready to get home. "If you ever need to make any extra money," she said as I got up. "We consider you family, you know."

I remember looking back at her as I closed the door and saying all I could think to say, "Thanks, Cindy. I appreciate that offer."

And just days later we were out again at dinner, with the money Sam would spend freely, the money that no doubt came from Cindy's body. Sam and Cindy suddenly felt like my apartment did—haunted, something I could not quite handle, though of a world not altogether unfamiliar to me.

At the same time I was starting to have more problems with my editors at the *Reader.* They were not liking my pitches and I was getting too many assignments I couldn't handle. I wasn't sure why I couldn't sleep or focus—at moments, I'd consider something was wrong with my health, but I still didn't have health insurance, so the idea of looking into it seemed like some distant absurd luxury. My thoughts remained on my dog's ailments. Kingsley was now needing to go to the vet weekly for more fruitless testing that revealed little more than Lyme. Jerry was apparently dating someone new. I was feeling more and more alone.

In my final months in Chicago I was a hair model to make extra money. I let the hairstylists at a salon in Wicker Park shave part of my head and I walked home trying to ignore the whistles of punk boys who called me *skinbird*. I was smoking again. I was feeling restless and wanting to quit my job. I was needing to get out of that space. I was low on money. I thought of Cindy's work offer and instead took on a nude modeling stint for a website and instantly regretted it.

I had started to take Xanax again for sleep, which either resulted in no sleep or sleeping all day. Something was wrong, so much was wrong. I started to feel like Kingsley and I could not do this anymore, that we needed saving fast.

So I did what I seldom did and I called my mother. At first I tried to calmly tell her things were not good.

"I have very little money. The job does not offer much. I have trouble writing. I take these pills . . . I can't sleep or I sleep too much. I have weird friends. The dog has been suffering here. I don't know why I am here anymore . . ."

Soon, though, it turned into me pleading with her to come get me and take me home. I knew this could be a terrible idea, but I knew that I needed a new idea, any idea. My life felt so unwholesome that a parental component at least felt safe. She obliged mainly because she had never heard me sound like that in my life.

At that point Los Angeles had only been the home of my childhood or the place I went home to for college breaks, not an adult home at all. I had no place there, but did I have any place anywhere else? New York was always my answer, but I couldn't imagine it now after Cameron, after 9/11, after grad school. And going back there with a dying dog and no job and this manuscript that needed so much work? No.

I realized going home to LA and not worrying about rent was the only option I had at this point. I made plans to take my yoga more seriously and to maybe take screenwriting classes, all side gigs I could develop while I bet on this novel business. I told myself I would save up and get back to New York again, but it would be New York with a plan.

The next stretch of disentangling myself from Chicago became torture. My LiveJournal friends were distraught, but no one more than Sam and Cindy, who promised to throw me a big party at their house. Jerry tried to stop me from going, but I decided to stop talking to Jerry. The *Reader* seemed perplexed at my move. My landlady was upset at me breaking the lease, but I let her know I knew what had happened in that space before me—even though I didn't exactly—and she didn't make another peep. I began putting Kingsley on a liquid diet so that he would lose five pounds he didn't really have to lose, because the only airline that would take big dogs in their cargo had a weight maximum—crate and dog could not top 100 pounds, and at current weight Kingsley plus the crate would be 105.

Like someone who knows their end is near, a period when they can afford to finally collapse and be done with it, I started feeling very ill myself in those weeks. I, too, was dropping weight, and I was constantly exhausted. It seemed likely that I was depressed, but I wasn't about to consider meds again. I would try to take runs to the Loop and back, and then I'd be sore for days, sleepless and feverish. I'd try again days later and it would happen again. Almost all food made me sick. I started to get worried, but then I consoled myself with the thought that my mother was coming, and she wouldn't let anything bad happen to me, and soon I'd be home in California. Somehow, everything would be solved.

My mother had no idea what she was coming to. When she finally arrived, both Kingsley and I were two thin restless ghosts. The apartment was a mess—I'd barely had the energy to clean up. She was most appalled by the state of my half-shaved head that was streaked in red—*I was the experimental hair model, Mom*—and I realized that this time was not all that different from when she came to get me from my year abroad in Oxford, where she faced the awful mess of my apartment and its ashtrays and my dyed hair, a result of punk and techno London club life. It was the first time I thought she knew I did street drugs, and a part of me wondered if she thought I was doing them again in Chicago.

*No one here does drugs*, I had told her. *My friend went to jail for it ages ago!*

I don't know how I felt so comfortable to tell her about Sam's past. Maybe because she was going to be staying in their spare bedroom, a room I had never been to, but that Sam immediately offered up when he heard my mother was coming to town. He was so excited at a Middle Eastern mother being in the house, but a part of me worried she was staying in the room Cindy ran her business out of. Or it could have been the room he stayed in with his gun. In either case, I worried about her being there, but she was relieved to be staying around the corner after seeing the condition of my apartment.

Sam and Cindy went all out at my going-away party. They invited everyone I knew plus some of the LiveJournal friends I'd never met. There were plates of hummus and baba ganoush and grape leaves and tons of alcohol, even though Sam could not drink.

My mother was the first to get drunk.

"It's such a beautiful home, don't you love their life!" she gushed.

I wanted to tell her, *If only you knew*. The bedroom she was staying in certainly had to be Cindy's lair, I decided, full of feminine satins and silks and even a nude portrait of Cindy on one wall.

"What's wrong? It's okay, they are married!" my mother tipsily laughed. "You need another drink clearly!"

I did. In the main living room area, Sam was blasting sophisticated blues rather than the punk or hip-hop he and I usually preferred, and all sorts of drunk people were coming up to tell me they would miss me and that they loved my mother. Finally one LiveJournal acquaintance asked if it was okay for us to do coke in a bathroom, as he had brought plenty.

My first instinct was to say absolutely not—that not only was my mother there, but that this was the home of an ex-con who'd done hard time for drugs.

But before I could say a word, there was Sam interjecting and telling the guy of course he could and reassuring me that they'd keep my mom busy in another space if people wanted to do it. Sam seemed totally unconcerned.

"Are you sure?" I whispered. "I mean . . ." I had known a lot of drug addicts by then and you didn't just casually open old doors like that.

He chuckled. "You act like I haven't done drugs."

"Yeah, but not recently!"

"Really?" his eyes flashed mischievously and without thinking I followed them to the bathroom, where both Sam and I did a line off a CD.

Nothing, not even the weirdness of knowing my mother was at the same party that I was at, could top the weirdness of doing drugs, even *just a little line* as Sam kept reassuring me, with someone you knew was sober, who had to be sober after nearly a decade of prison. My anxiety about it all made the little line I did sort of fruitless.

By the end of the night my mother and I both drunkenly stumbled home to a hard-panting Kingsley. My mother, never a fan of dogs, suddenly hugged him. I gave her her stuff as Cindy waited outside in the car to drive her back.

"What wonderful friends you have! Why would you want to leave this great Chicago life!" she said, laughing.

I woke up the next morning next to a guy from LiveJournal who had come back later that night. I kicked him out, gathered my bags, and met my mother. Our flight was that evening but there was much work to be done. We could either do it, or we could do as Sam suggested and meet him for a final dinner out, on him, *his harem!*

In the end I left the apartment as it was, still half full of my possessions. I stared at the mess for hours on end, unable to move. For some reason, I couldn't bring myself to touch my life there, to physically grapple with the pieces I had barely even put together. It all felt cursed. I imagined what it would be like to just walk out on your life intact and I realized here was my chance to abandon everything like that. What could I possibly want to still hold on to? I called Cindy and told her the door was unlocked, and that they could do as they wished with it all.

That night we joined Sam out for a dinner in the Gold Coast. It was the usual Sam, full of charm and flirtation, overwhelming the waitress with all sorts of fussy tweaks to a chicken dish, telling elaborate stories, and at the end, laying out wads and wads of cash, too much for one waitress's tip. "Don't even think of it—you are family!" he snapped at my mother, as she reached for her wallet.

I'm glad I went because it was the last time I saw Sam. He

ended up dying several months later. His heart had stopped. His brother told me he had been suddenly back on drugs. They had done what they could, but nothing seemed to stop him.

Many years later Sam and Cindy came up in conversation, and my mother was remembering them and her Chicago trip with me fondly. I told her he was dead and that it was drugs. I told her furthermore he had been to prison for many years for hard drugs.

"You would never guess," my mother said. "He was a writer like you."

I even told her Cindy was a prostitute. That surprised her less. "There was something about that bedroom," she told me. "Who would have a portrait like that of his wife, all naked?"

At one appointment with a doctor years later, one of the few I convinced her to go to with me, when the recounting of my physical and mental history became too much to bear, my mother interjected that she thought I got sick in Chicago first, saying I was never quite the same after Chicago. I was shocked to hear it, but she seemed convinced.

"There was something so wrong with her life, something so off with it," she said, "that I just knew we had to get out of there, and I couldn't seem afraid myself. But I know for sure she was sick."

# ON SUPPORT

At a talk I gave in Pittsburgh in April 2017, a young woman asked me afterward: "What was the worst part? Worse than the pain even? What odds were most against you? I don't mean being a woman, or a woman of color, or things like that, but what were the most personal things that you had against you?"

Before I could argue those very elements of identity she mentioned *were* personal, I realized she was far from done.

"You see, I'm very ill and at dead ends," she said to me, tearing up. "I have big things going against me. My family for one. Money for two. Sometimes it's worse than the pain actually."

I nodded. And nodded.

Anyone who has ever had chronic illness knows people will speak of your "support system." Family, friends. I had the latter, but the former was questionable at times. I only approach it as much as I can here—as much as I did to this young woman also—because of my culture. I am Iranian American and my parents, while old, are still living. I am conscious of the fact that I am not telling their whole story—including some of the less forgivable ways they dealt with my illness. But I want this space to exist here, too, this absence, that for the most part my family was not always part of my support system.

Not my chosen one at least. A dozen friends, which later became a half dozen, which is now a handful at best, became the real family. Some of them have Lyme, some of them have only heard of it through me. But they know it inside out. They hear me, they believe me.

And the deal with so many chronic illnesses is that most people won't want to believe you. They will tell you that you look great, that it might be in your head only, that it is likely stress, that everything will be okay. None of these are the right things to say to someone whose entire existence is a fairly consistent torture of the body and mind. They say it because they are well-intentioned usually, because they wish you the best, but they also say it because you make them uncomfortable. Your existence is evidence of death, and no one needs to keep seeing that—especially not the people who gave birth to you. They are not supposed to live to see you not live. It's not the way nature is supposed to work, after all. (And yet nature looks that way all the time—it's just our morality transposed on nature that requires it be benevolent.)

Only at times, might they really see you. When my mother admitted she saw me as sick, I cried for hours. She thought she had done something wrong, but I let her know it was the opposite. To be seen, to be heard, to exist wholly, whether in beauty or ugliness, by a parent often felt like another big step to wellness. I experienced it rarely, but when I did, I felt something light in me that I had long thought had burnt out.

I think in the case of my parents shame about money came into the picture often too. If my parents could not support me, how could they support a monster illness? They were of an immigrant group that was supposed to be bringing riches to the

United States with them after all. But this was not them. They watched me spend every penny I'd saved on figuring out what was wrong with me—often to no end. And even when I found out what it was and figured out how to crowdfund the costs—the only way I was able to pay for this impossibly expensive journey—my public calls for donations embarrassed them. At times, friends have sent me money and I know they know this. I avoid crowdfunding these days, trying hard to stay afloat by working many jobs—though I know that working many jobs is what makes me sicker—and they often imagine I am now well, even as I go in and out of hospitals, even though I often walk with a cane and use wheelchairs at airports.

This was not the life they imagined. They were supposed to have money, I was supposed to have health, and all of that was supposed to be tied up in the same bundle. Health and wealth. I think the only thing that consoles them is the fact that it seems like a big chunk of Americans are also without those things. All my life, I've heard my parents and relatives say *America is a sick country,* in every meaning of the sentence possible.

# 4

## LOS ANGELES

I can't recall a time when I felt my parents' home was my home. Part of it was their own fault—I remember being single digits and echoing my father's constant refrain that we were in America only temporarily, that we'd be back in Iran soon again. The revolution, the war, the Islamic Republic, it would all pass soon, and then we'd be there—*home*. But it never came. It still hasn't come. My parents bought their first property—a two-bedroom condominium in Glendale—in 2011, after more than thirty years of renting.

My parents always called me a wanderer and poked fun at my comfort with a sort of homelessness, while also realizing my wanderlust was something chronic, less leisure than discomfort. It dismayed them, and they blamed themselves, pinning it on the turbulence of my first few years of life in escape and exile.

It had always been clear to me though that Los Angeles wouldn't work out for me. My New York friends found it so fortunate that I'd work in New York and then go home to Los Angeles—*like celebrities*, my mother would remind me. But Los

Angeles never felt glamorous. It felt dark, dangerous, lonely, full of trauma and isolation—all the excitement and uncertainty and mania were replaced by the endless-summer depression, the bleakness of little potential, and the unchanging lull of suburbias layered upon suburbias.

I had many demons in LA. My childhood with ever-fighting parents had been an unhappy one, and I'd lock myself up to write on the weekends, not out of a joy for writing but out of a desperate will to escape. There was a constant embarrassment of poverty in a city where many had some sort of handle on wealth, not to mention the suspicions and complaints directed at my family for being foreign and from a place that others most likely weren't supposed to like.

Still, at the end of 2004, after my troubles in Baltimore and Chicago, I was full of hope that LA would work. There was screenwriting. There was yoga. And I'd go to the beach more, and not dressed in all black the way I did as a teenager, when I was rebelling against all that was gold and healthy and tan and bare.

At yoga teacher's training, I tried to remain grateful for the scholarship that had me affording the elite Santa Monica yoga school, where we were told not to eat a thing for six-hour sessions. I resorted to sneaking granola bars in the bathroom. While I had loved yoga since I was eighteen, I hated yoga school. I found it punishing and petty, the spirituality fake and exhausting, the emphasis on bodily perfection alarming and backward. Whenever I was praised for my form, it was evident that I was being praised for being thin. And it was there that I started going through problems with my savasana, or corpse pose. Every time we did savasana, I'd fall into that desired strange half-sleep state that the yogis speak of as optimal. But without fail I'd be yanked from it by a violent shock. At first I likened it to falling dreams, but it soon started to feel like convulsions. I began to fear savasana and tried to go to the bathroom until one of our instructors, Lisa, noticed.

"What's been going on?"

"With what?" I tried to act casual.

"You aren't quite staying with your savasana," was how she put it.

"I can't handle it," I replied. "Something about it has started to feel wrong."

She asked me to try to describe it and then to try to pinpoint what was wrong.

I told her my body felt like it would go into shocks, and she said something vaguely metaphoric about a snake in the spine and "the kundalini," but I barely listened.

After that I tried to really examine the feeling, understand why it only announced itself during savasana, and I began to notice it was something that happened when my body would calm down after intense exercise. It seemed my body was having a rough transition to repose, almost like it could not slow itself down. My heart rate in particular seemed to have trouble readjusting to rest.

I tried not to think about snakes in my spine too much, but there was something off with my body that was not so subtly announcing itself. The moment I'd start to consider that something might be wrong with me, though, I'd shake off the thought. I had struggled enough in recent years. It was time to divorce myself from troubles, from even the possibility of troubles.

In my quest to ignore these issues, I tried to chip away at that first novel manuscript, sending updates to old writing mentors here and there, asking if they knew of jobs in New York, always wanting to keep that door open, even if just a crack.

I knew I was buying time, that I was in some sort of midlife downtime, and that something was bound to happen and life would surface again.

I took an extra job as a shopgirl on Rodeo Drive—an undergrad at Hopkins had a mom who was a Palm Beach luxury handbag designer and she wanted to set up a Beverly Hills outpost. I'd been a vegetarian most of my life and the entire store was skins, from alligator to crocodile to snake to ostrich. I went to work with more dread than even my worst phases in school, my body barely able to drag my psyche into that shop in the part of Los Angeles I disliked most. I used to tell customers—usually older Russian or Korean women—to come next week when we'd be having a huge sale, because I had too much anxiety to ring up the merchandise—sometimes upwards of $50,000 for a bag. They'd never come back. I made nearly no sales my entire time there, and no one seemed bothered by this.

One Tuesday morning, when I went early to open the shop for the cleaners, I received a call from a 212 number on my phone. Knowing it was a Manhattan number spiked my excitement, and I answered with much more energy than I typically felt at 7 a.m.

"Hi there, Porochista? This is Elisabeth Schmitz, an editor at Grove Atlantic."

I froze. I suddenly remembered an email I had sent months and months ago, when I had reconnected with an old friend from my freshman-year workshop at Sarah Lawrence College. He was leaving his job as an editorial assistant at Grove, and on his last day he asked if he could put my novel draft in the slush pile just for the hell of it.

"I'm not done yet!" I told him.

"You have a draft, no?"

"Yeah, I had a draft in my fellowship year after grad school. But I've been working on it and it's just not nearly done."

"What's the harm?"

"I'm not in a rush. I'm not a child actor." I felt mature in my response.

But in the end I caved, and now here it was: a call from my dream editor, who had happened on that very manuscript—barely spell-checked.

"I hope I haven't reached you at a bad time—I realize it's early—but I wanted to reach you fast: I have been reading your manuscript on the subway, at work, every chance I get, just laughing and crying my way through it . . ."

It felt like a movie: an editor going on about my novel while I sat in that desolate shop, clad in the uniform black dress and leaning into a broom like a modern literary Cinderella, as the cleaners dusted every skin-purse display around me.

"I was wondering if you were ready to go forward."

I knew the correct answer was no. That I was not done with the manuscript. That order of operations dictated I get an agent first. She seemed to agree.

"But can I come meet you?" I asked. Every part of me wanted to be in New York, where things happened, unlike LA, where I felt dried up and stalled.

Things were set into motion with that call, and over the next few weeks I began to furiously edit my manuscript and distance myself from yoga—I got a C on my teacher's training final and I walked out perfectly content to never take a lesson there again. I began sending letters to agents and I started to hear back. I was focused on imagining a life in New York again. It was happening.

---

It was an odd feeling visiting New York without living there. I had agent meetings. I realized in LA I was always in either a black cocktail dress for shopgirl life or yoga pants and sports bras for yoga school. So I had bought a bunch of what I thought were writerly looking Banana Republic separates on sale, and I was relieved the half-shaven Chicago model hair had grown out to what people might call "a sensible bob." But I wasn't sure what role I was being cast in. All my life I thought I knew how writers looked—like me—and now that I was about to be one, I had no idea.

Alexander was eager to cement himself back into my life, as he'd felt I was slipping. We were staying with his parents in Union Square for the visit, and he was so proud to show me his parents' house. I couldn't wrap my head around the fact that he, like some Henry James protagonist, had just grown up in the middle of the city, in Union Square. The apartment was huge and impeccable, like the ones you saw in movies, with their own elevators, full of paintings by major artists you could never name, and tapestries and rugs from around the world that probably cost as much as a car. Alexander's mom was a yoga obsessive—my yoga training was the fact about me she was most interested in, as much as Alexander tried to emphasize *writer*—and she owned a (*slightly better than mediocre*, Alexander would call it) restaurant a few blocks away. Her new husband (not Alexander's father, who was the source of the old family money and who lived in the Hamptons) was responsible for many famous paintings at the Museum of Natural History, a curmudgeonly guy who more than once made me cry.

Their apartment was not a comfortable place for me—wealthy people in New York always made me feel out of place—but it felt somehow auspicious as a headquarters for agent finding. The biggest problem for me was their giant parrot that would roam the living room, a loud, huge, ancient, gawking beast, who'd unpredictably land on things, even you, as he pleased. I hated that bird and couldn't even pretend to like him. The only times I actually imagined being married to Alexander were when I imagined that bird outliving the both of us.

I went from meeting to meeting and all went well enough—I knew I was someone who could cope well in meetings, whose meetings would surpass the awkwardness of my query letter posturing. My final was with an agent who was then known as the hottest in the city—a young, handsome, somewhat enigmatic agent who had recently discovered one of the foremost experimental writers around. We had a long lunch at a chic Chelsea bistro and he seemed interested in other projects beyond my novel, projects that would have nothing to do with my Iranianness. He ended up being my choice, and as I left New York I asked him to confirm, more than once, if I really and truly had him as my agent:

"Yes, I really and truly am yours. Congratulations. It's happening."

Alexander took me out for a fancy meal to celebrate, but he was already disappearing to me. I chatted on about the months of edits I'd have to do, and then my dream: the inevitable sale of my novel, the potential auction, and then my glorious return to New York City, with a real life this time.

"We will be the happiest couple in the city," he chirped, sipping his wine, but I could not understand what *we* or *couple* he was talking about.

---

In the end Alexander left for New York without me, before me. He and I played a messy tug-of-war between New York and LA; he the New Yorker, I the Angeleno, with our work lives requiring that we often had to be in opposite cities.

Maybe he had been quietly looking to leave as we readjusted

to our life back in Los Angeles, after that whirlwind New York trip. Maybe it just happened by accident. But when he got an offer to be a showrunner on a show based in New York, he took it, and assumed his departure was a logical opportunity for me to go with him. I resisted.

"What do you mean? You always wanted to go back. And you need to go back now."

"I came here for yoga. For screenwriting." I felt myself hovering around the lie. "I don't have a life set up there yet. I have to work."

"Come on, we can get back to your old life there, that's all you ever talked about. How LA is never home."

"It isn't," I said. "But it's all I got." I didn't know what I meant.

It took only a few months of him being there for us to break up. He sensed it and he made the final move—*I think it's clear to me that the best thing I can do for your life is to leave it*, he said through tears on the phone—but I was done months before that. I felt his loss so slightly, a hangnail, a ringing in my ear, a twitch. Just like that I was back to my element: alone.

As I began to work with my agent remotely on the novel, I also had one foot in with a Middle Eastern luxury and high-fashion magazine. Something new. Funding from a Kuwaiti sheik. Huge budgets. Easy hours. I could be the features director.

I hesitated. It was a moment in my life when I felt it was important for things to be about me. My novel, my career. I was tired of supporting the work of others as I'd done for years as an arts and entertainment journalist. I tried to politely decline, but I knew it was not the last I would hear about the project. The editor was hard to say no to.

Less than two months later, my new agent and I parted ways, as he seemed to have no time to answer an email, much less work with me on a manuscript. I tried to be confident that things could happen with a new agent as easily as they did the first time, even though I would be searching remotely. I sent out a query batch to new agents as well as some from my previous outreach.

This time: nothing.

I started to panic. Was I back where I started, but worse?

I wrote the Grove editor who had loved the book to tell her of my problems, and she agreed it would probably be easier if I was in New York.

I thought of Alexander in that giant house with that awful parrot, who had said, even as we broke up, that he'd always be waiting for me.

I told myself that now that I had my yoga teacher's training certificate, I could teach yoga there. And if not that, I'd find other jobs.

I called the magazine editors and told them I'd try out the role of features director of this magazine, but that hopefully it would not be forever because I was coming to New York with real goals—my one goal, all along, my whole.

I was coming back to New York to be a novelist.

---

Living in New York for nine months, arriving in fall 2005 and staying until spring 2006 was like old times: just another school year. Life was supposed to be especially neat: get a new agent, get a book deal, and figure the next part out. Instead, other things found me. Julie, my old Sarah Lawrence friend, and I wandered around the East Village, talked of old times, and pretended to relish in the blank slate of our futures. During the day, I'd also meet with my magazine editor, Samir, in his Chelsea apartment, the luxury magazine *Alef*'s headquarters, where we dreamed up all sorts of impractical fantasies that we knew could actually likely be fulfilled thanks to our sheik's millions (if not billions).

One night my editor and some of my old *Paper* magazine friends went out to a club—it was as usual a bunch of gay men and me—and Samir had dressed me up in clothes from a vintage clothing store he co-owned: an impossibly tight black suede eighties YSL hobble skirt, and a black sheer Gucci blouse with strappy Jimmy Choo black stilettos. As they dropped me off on my corner, I—constricted by the hobble skirt and the high heels—drunkenly tipped into a man waiting at the crosswalk. He was a handsome bald guy in a pale suit and I immediately recognized him: the writer of the moment, the writer everyone was talking about.

I blurted out that I loved his writing and that a Chicago ex-boyfriend of mine had recently interviewed him.

He seemed delighted by it all. "Are you walking this way? Taking a walk maybe?"

It was late, but he insisted I come along on his walk to St. Mark's Bookshop to sign some stock before it closed, which I found a very glamorous element of what writers did.

The night was one of those balmy perfect end-of-summer nights, and I soon realized The Writer could not take his eyes off the sheerness of my blouse or tightness of my skirt. I tried to explain it was not my usual style. I watched him sign some books at my beloved bookstore, the staff delighted to see him. After that he suggested a drink, though he quickly told me he did not drink. I realized the purpose of the drink was to get me drunk and I happily obliged—I could handle this—sangria after sangria at a strange desolate bar in the East Village. It was only a matter of time before he grabbed my face and began kissing me, as if it were a scripted destiny, and I was in awe of him and how great it all felt. He walked me home, and we planned on dinner the day after next.

In the morning, Julie and I gossiped for hours about my older man, my famous man, my writer. We got smoothies and went to the bookstore and skimmed his books for the dirtiest details. Julie, a stylist, planned my outfit for the next night.

But that evening, she and I had plans of our own—another collegemate's party in Brooklyn. So many of the rich Sarah Lawrence girls had turned their trust funds into brownstones in snugly gentrified Brooklyn hamlets and discovered a sort of suburban domesticity I had felt dedicated to avoiding.

I remember staying longer than Julie and most of our friends, leaving much later than I had intended. I remember that weekend there had been a subway terror alert, and so we were all taking cabs. I remember my brown fishnets, Charlotte Ronson platforms, trench coat, and paisley silk dress with the velvet ribbon belt, cobbled from sales and eBay. I remember finishing my second or third red cup full of mediocre merlot, and asking where the cabs in Carroll Gardens were, and someone I didn't know well telling me Smith Street, and me instead walking down to Court Street and catching a cab immediately. I remember fiddling with my purse, as the driver sped through the misty, London-esque drizzly night from Brooklyn to Manhattan. At some point I recall looking up and catching the awning of an old fried vegetarian place I used to love, Spring Street Natural, back when I was a vegetarian and somehow must have had enough money to eat out. The next thing I remember was a hard crash from the back and another impact, and the glass of the cab shattering, and my body flashing in pain from several points, and blood on my hands from my face. And then my own animal scream, my immediate impulse to run out from the car, which had become a warped twisted thing, and running into a crowd of people, mostly men, all smelling like smoke and alcohol after a long night out. I think the club Butter was right there, but it was a man who I later knew as a waiter from nearby Savoy who took charge and told me it would be okay, and who refused to show me my reflection when I demanded it.

In the ambulance they looked at me with concerned eyes and gave me oxygen, and I kept talking and talking, because I had read somewhere that's what you had to do not to fall into shock. I didn't know exactly what the problem with shock was, just that you had to do everything you could not to fall into it. I remember asking the paramedic what would happen since I had no health insurance, and he told me he had a good feeling I wouldn't be paying for any of this. I asked him if he thought I was okay, if I was going to make it, and he told me that they could never say that to injured people for sure, but that he thought so anyway.

At the ICU in St. Vincent's I kept asking for my purse and compact, knowing something was very wrong with my face, knowing that though my ribs and shoulder felt off, my face was where the bulk of the damage was. I landed on my cell phone and pretended to call some worried party and instead took what today would be called a "selfie," using the camera as a mirror to see if part of my face was missing.

And it was—my photo revealed that a portion of my face had been torn off, though everything was covered in so much broken glass that it was hard to evaluate the wreckage of car and skin all intertwined. They quickly transferred me to burn trauma and the surgeon sewed plastics onto my face. I left the ER that day with my face wrapped in gauze, how it would be for weeks.

An old professor picked me up and we sat in her living room, both stunned. I worried about the date I was supposed to have that night, my upcoming date with The Writer. My professor insisted I call it off instead of sitting and worrying about it. I was going to be out of commission for a while, she said. I thought of my plan to support myself by teaching yoga and realized a lot was about to change.

When I called to reschedule, The Writer said of course he understood, though he seemed startled in that slightly awkward way one might be when a near stranger has very bad news. He tried to offer to help but then commented on the strangeness of it.

"I mean, I'm sure you have people around you who can do it, but if they can't, I'm here."

"Thank you. I'm sure I will be fine in no time. And then we can see each other!" I tried to sound cheerful but realized it could sound crazy, given that I'd just emerged from a major car accident hours before.

"I'm sure," he said. "One day. No rush. But for now, you have to take care of yourself."

I heard that over and over, *take care of yourself*—a shabby, worn mantra that haunted me through endless hours in bed in the apartment for the next few months. I couldn't tell at times if I was depressed or just "taking care of myself." Julie would bring me the CVS ramen and offer to make me something more, but I'd refuse, wanting to hold on to something normal. I spent several days a week at physical therapy, telling the therapist that my goal was not to just be better but to go back to teaching yoga.

He was suspicious of my form. At several points, he wondered if my shoulder problems weren't more casualties of ashtanga yoga than a car accident. At another point, as he fiddled with an electrical stimulator, I felt confused by my reflexes.

"Are you sure you don't have any other conditions?"

Conditions? I nearly laughed it off. I had no time for conditions.

"No. What's wrong?"

He shook his head, baffled. "It's hard to say—it's just intuition. Something sometimes seems off with you, with your muscles and nerves, and I have to wonder."

I rolled my eyes. "No, you don't," I said. "No, I don't."

*No more troubles* was always what I was thinking. A part of me suspected he was right—all sorts of things had felt off in the last few years, but I never gave myself space to suspect them. At night, I either couldn't sleep or I'd fall into something like savasana's corpse pose falling dream electricity. I was full of aches and pains that seemed to travel around my body, and I often felt feverish but refused to buy a thermometer to know for sure. The odds of a new problem, another thing, seemed just distant enough for me to ignore the possibility. But I was conscious that I was ignoring something. At times, I watched my fingers grow numb and tingling at the computer keys, and I tried to merely marvel at the phenomenon, rather than be shaken by it. I couldn't afford to know. The most I'd do was reach for a Xanax or two, the pills now positioned like good friends—what I had once been prescribed by the Johns Hopkins psychiatrist now was a daily prescription by my own psychiatrist aunt. The Xanax never made the pain go away, but it made my reaction to pain go away.

For days I could watch myself just not caring, a sort of dazzling indifference, a mute button almost creating a lovely white noise, its antipresence so very present like another hole in my life.

---

Within a couple months, The Writer and I were on that long-awaited first date, my magazine editor lending me another YSL number for the occasion, a black corseted cocktail dress.

I made sure The Writer was sitting on my left side, since the right side of my face was still fairly immobile, much like a stroke victim's, people said. The scars had left a constellation of puckered skin and for once I was almost grateful for the adult acne of my twenties that helped them blend in. It was mostly my mother who was horrified by my face being disfigured—my mother and my attorney, whose entire case seemed to hinge on the fact that I was a woman of "marrying age," twenty-seven, and now I had a heavily scarred face. I kept reminding myself how lucky I was, how easily my brain could have been damaged, how close to losing my one and only real worth I had come.

The Writer told me over and over I was beautiful, and we dated for months following that, as he came in and out of town on book tour. He never quite committed to me, though he was never absent for too long either. Then the greatest gift from The Writer came in an unexpected way. I had made little progress with agents, and he told me one night at dinner that if I wished I could always write to his agent. I prepared a package for her and he sent it off. In a matter of weeks, she came from her home in Princeton to meet me at a fancy restaurant in the Village with a contract in hand. She loved my novel.

And like that I had my second agent. And a first novel that was being shopped around. It was not the narrative I had dreamed of, but it was a narrative I knew others would see as well.

---

By the time the book sold in the late spring of 2006, so much was different. The Writer was gone from my life—he could not commit and I lost interest in all his impossibilities. Alexander and I had reconnected, and we gave our relationship another chance. I had decided to leave the magazine, which had been far too stressful while also never really paying much. I was no longer going to physical therapy and my face had mostly healed, except for one neat slash that has permanently warped my smile. I had the personal injury lawsuit in the works, though it would turn out to be years until its closure and payoff. I was also on my way back to California for the summer, as my room at Julie's had only been a temporary solution.

I was heading back to California with the intent of coming right back to New York in the fall, just two months later. I was dedicated to completing some edits that my new editor, Amy Hundley of Grove Atlantic, had suggested while I was there. (Amy was a trusted colleague of Elisabeth Schmitz's, whose list had been too full to take on my book in the end.) The Xanax was still providing me almost instant relief from stressors in my life; it was like being dipped in a cool pool on a hot summer day, no memory of the agony and sweat that had sparked the taking of the pill. But I was going home as a success—an author, a person who had made it. I told myself that this had to be the beginning of the end of my anxiety and depression.

On the flight home, I was a wreck from turbulence, so I popped extra Xanaxes. But when I got home, I looked at the many bottles of Xanax I seemed to have in my possession—it looked more excessive than I could justify—and told myself I would not keep taking them. After all, what did I have to be so anxious about anymore? I was on my way to having it all, or so I would have to keep telling myself: boyfriend, agent, editor, book deal, recovery from a terrible accident. Absolutely everything for once was fine. It had to be fine.

And for a few days things actually were fine. My parents were happy to see me, I went to yoga class, I cleaned out my desk area, and I set up a writing schedule—one that revolved around intense focus, breaks, excellent nutrition, yoga, and plenty of sleep. Though it was that final item that proved to be an issue. I'd never been the best sleeper, certainly not as a child, but I blamed the last year's poor quality of sleep on the car accident plus just being in New York City, which was not a place anyone would imagine would yield great sleep. My aunt had warned me against taking the Xanax for sleep—she claimed it would compromise the depth of REM sleep, plus that was where they found people had the most addiction issues, when it was tied to their sleep—and so that was not an option. Not yet.

But it turned out that a few days into my being back, I seemed to have altogether forgotten how to sleep. Years later, Dr. E, my infectious disease specialist, would tell me that this made complete sense—that these were textbook symptoms of Lyme. That you hear a lot about the joint pain and muscle aches, but that the first symptoms of Lyme for most are anxiety, depression, and insomnia. The sort of insomnia that does not let up. The sort of insomnia that does not respond to conventional medication. Lyme attacks soft tissues in the body, and the brain is one of its favorite organs to feast on. Almost always, as the spirochetes (the spiral-shaped bacteria that cause diseases like Lyme and syphilis) multiply and infiltrate the body, the Lyme sufferer loses the ability to sleep. *It's usually a particular type of insomnia*, he said, *the kind that really ruins people. It's not the type of insomnia the general public can begin to fathom.*

I remember the first night of it well—that old hot summer of my California childhood announcing itself in June, the beginning of its season of unrelenting presence. I remember wondering what it was that was keeping me up, why I wasn't tired, was it something I had eaten, was it something I was worrying about, was it something wrong with my body, was it something happening to me physically? I couldn't understand—I'd never had insomnia so absolute. I ended up staying awake the whole night—something I hadn't done since I was an undergrad—and skipping yoga in the morning, too exhausted to imagine it, but also feeling like the sleeplessness meant some sort of event, and that now my day should revolve around its absence.

I didn't worry. It was a one-off. But all day I remained vigilant for clues as to its cause. I started to notice we had new neighbors upstairs and you could hear a lot of stomping and music playing. It was perfectly possibly that their rhythms had kept me up. I decided I'd buy earplugs and, if that didn't work, white noise. There were options. But I also decided that it might not be a big deal at all and everything could be fine the night ahead.

Except that it wasn't. I put the Rite Aid earplugs in at 10 p.m. only to yank them out at 2 a.m., after several hours of more insomnia. How was I not sleeping? I had resisted napping all day, just so I'd be ready for a full, long, luscious night's sleep. And now, what was going on? Again, I scanned my body and my surroundings and could not pinpoint the catalyst. It had to be my mind. But why? What was keeping me from sleeping?

By the end of that week, after several nights of dramatically disrupted sleep—I'd eventually catch a couple hours here and there but it was nowhere near a normal night of sleep—I decided that maybe a good use of my advance money was therapy. I had no health insurance and, outside of no-fault insurance, I barely was even covered in my twenties and thirties. Alexander had recommended cognitive behavioral therapy to me before, as it was something his whole family did. I searched for a CBT practitioner near me and found one—I noticed on her website "insomnia" was a prominent category, one she linked to depression. But was that it? Would depression mean essentially forgetting how to sleep?

She seemed to think so, and after a few sessions, as my acute

June insomnia turned into July insomnia, she was convinced I had depression. I told her it was the happiest time of my life—I had a book deal, my life's dream—but she noted it was a major life change and all sorts of major life changes could cause depression.

I was skeptical, but I took all her directives and became a person with a journal again, someone in the business of tracking my every mood and anxiety. I tried meditation and breathing exercises at night. I tried to get up, I tried to sit down. I tried to read, I tried to watch a movie. I tried to eat something. I tried to be relaxed about it, not to panic. I tried seemingly everything, but daylight would always win just as I felt I was warming up to the idea of sleep.

I talked to my CBT practitioner about the Xanax, imagining it could help, but she seemed wary and referred me to a colleague of hers who was a psychiatrist. He handled the drugs. But she kept emphasizing that with CBT they liked to use that as a last resort.

"Well, I think it's time for last resorts," I remember saying. "I really haven't slept in a month. Any sleep I've gotten has been an accident, it feels like."

I remember she nodded very slowly, as if in defeat.

"I mean, I have a book to finish—that's why I am here," I said. "I wanted to return to New York in another month, but I've gotten nothing done. I want to get to my life."

I found myself saying the exact same thing the next week to Dr. Toll, her colleague, who watched me with an anxious look in his eyes.

"Why do you think this is happening?" he asked me.

I told him I had no idea, that I had tried to think of everything external.

We both sat in silence for a while.

"I mean, I guess, one should not rule out why I am here."

He looked interested suddenly.

"It's possible I've completely lost my mind," I said, plainly, and it shocked me how sensible that sounded to me, if it weren't for the fact that my voice didn't even seem like my own.

---

That summer of madness, a lifetime or more. Days went by impossibly slowly. I accepted that I was mentally ill, deeply incurably so, and this was to be my life. Still, at times I fought it. I became someone whose main job was trying out medications and going to the doctor. It was like shopping in a way: I found myself spending a full day googling and consulting friends on Seroquel, an antipsychotic I'd been prescribed—which back then was fairly new and had all sorts of black box warnings—only to refuse it in the end. I tried out antidepressants with names like sci-fi wizard goddesses, Paxil and Celexa, and always felt moments later that my entire body was burning, quickly discovering that I was someone who came down with severe neuropathy when it came to SSRIs. I tried every natural supplement you could find in a Whole Foods. I tried acupuncture, I tried an ayurvedic center, I tried multiple healers, I tried nutritionists. At one point I was seeing three different sleep specialists who all seemed fairly invested in hiding how stumped they felt. I spent every penny I had searching for the energy to keep seeking.

*I think something is wrong with me physically*, I caught myself saying again and again.

At a certain point, they all gave in to that logic. Of course not sleeping would harm you physically—it was a fact. Even if I didn't start with a medical problem, I now had a serious one on my hands.

Meanwhile, no one knew what it was that was physically wrong exactly, just as they could not pinpoint what was wrong mentally. My anxiety and depression seemed to scissor each other in ways that mimicked bipolarity. Insomnia was a component of both. One doctor gave me the diagnosis "somatization disorder"—*it's basically like hypochondria, but with true symptoms, only we can't find their physical source*, a doctor boiled it down for me—and then told me he was quite reluctant about the label as there were few doctors who knew just how to treat it. It seemed hopeless.

Eventually I was given an Ambien prescription, what all my doctors agreed was a temporary but proper solution. I remember feeling excited by it, as it seemed like something that was fashionable to take at the time, its use was so widespread. Feelings of hope could still bubble up in me at times, even as I spent most of my days feeling dead inside, desperation robbing me of any positivity, any trust, any way to see past this. But then every time a doctor had a new plan, I'd fall into it headfirst, as if enlisting every last muscle and nerve to do its best.

Something had to work. Or else.

The threat I had for myself was always dangling somewhere in the distance.

Something, anything.

That first Ambien night was a milestone for me—the first time a pill had completely solved a problem. I took the pill and waited in the dark like a child for Santa, and before I knew it it was the next morning, and the first time I had gotten eight hours of sleep in weeks.

I wrote my best friend: It has happened. Hope! I actually slept. Maybe this is a new beginning!

At the beginning I worshipped Ambien. I could not wait to get to it, that nightly ritual of having everything in place, the chamomile tea, the spray of lavender, the ear plugs and satin eyemask, but knowing they were all just supporting cast for my real star, Ambien. Pill swallowed, the curtain dropped, and the next morning debuted. I loved the feeling of control it gave me. I felt I had a solution for life.

Until it stopped working. One doctor had warned me it would be a short-lived solution, eventually only working for half the time. And like clockwork, after a couple weeks the eight hours abruptly became four.

I wrote friends: I don't know what to say. Let's delay the congrats. I'm back to bad sleep. It seems to have stopped working. I don't know what my options are now. I don't even know if I have any.

My main sleep specialist seemed unconcerned and said I could take a second pill after the four hours. Or I could do what I'd originally kept asking about—I could take a Xanax on top of it. I questioned the mixing of the two, but my doctor laughed at my apprehension. "People are mixing things all the time, and far worse things," he said. "You'll be fine."

And for a while I was close to being fine. I'd take the Ambien, wake up four hours later, turn to my bedside meds bowl, and take a Xanax, and then I'd get that extra four hours of sleep that followed too.

It was around this time that I began getting panic attacks again. They came like clockwork, always around the same time, close to noon, always a few hours after I'd risen. Something was terribly wrong, and with daily panic attacks now in the picture, I began to find that the daytime rivaled the night in terms of horror. I began to sleep more thanks to an increasingly complex regimen of pill dosing. I began begging my mother to sleep on the floor in my room, or my brother when he was home, or friends I had in town.

I was months into my LA stay, and it was about the time

when I had originally anticipated going back to NYC, supposedly with a fully edited manuscript for my editor to begin the process of having a book launch in a year. But I had nothing. I had only the beginnings of an illness and a breakdown, and I was losing the battle of managing them. *Chronic fatigue*, I told my editor, *that's what they are saying*. But no one was saying it and there was no "they." But I had looked it up and it didn't seem impossible.

The truth was, I didn't know what was happening. It seemed like I could not remember how to sleep anymore. I'd try to imagine an elevator dropping, I'd try to imagine falling off a cloud into more cloud, I'd try to imagine curtains dropping, I'd try to imagine that other world of dreams was the better world, but it was no use. And it seemed the close relative of insomnia was a panic disorder, the clockwork shakes and shimmies that would rattle my frame pretty much every day. I started to think it was my system withdrawing from the sleep meds rapidly, their half-lives going to a zero, and my body protesting.

I also couldn't help but wonder if something else was going on. Was this really the pills? My injured brain? Was something off on a deeper level? But the pills didn't give me a break to explore it, and I was either trying to fall asleep half the day or cope with being awake and enduring the suffering that was consciousness. I remember realizing at points that what was missing each day was quite simply living.

And of course, a third element then came to visit, less Technicolor than anxiety and panic, but a dark blanket of smoke, unchanging, unbreathable, thick, immovable: depression. And with all these illnesses my body was failing—I was losing weight rapidly in five-pound chunks due to chronic diarrhea, my hands would shake so hard I could barely hold a glass of water, and I was so fatigued that more than a walk from bedroom to bathroom meant an hour of lying down, unsleeping, staring blankly at the ceiling. Everything hurt, every part of my body felt like its wiring was all wrong, I felt like a foreigner in a hostile country, never adjusting or accepting that this was what it had all come to. I couldn't quite fight it, but I could not be at peace with it either.

By then I had become a regular at the ER, and I was past the point of just relying on Xanax as my benzo. I had encountered Ativan, which was supposed to be the safest, the nurse reluctantly told me, and the fastest acting. Ativan, plus Xanax, plus Ambien would sometimes allow me to cobble together a night's sleep or something resembling one, each drug doing its shift for a couple hours and then passing the baton on to the next. That also meant I'd have to do some benzo a few hours after I'd wake up. For a while, I didn't let myself think too hard about the dependence that had become an addiction at this point, and how I needed to keep some drug in my system now during the day if I wanted to survive the night.

I started to dream more and more of suicide. I knew the story could not end well. The narrative had taken an unexpected downward turn—the girl who finally achieved her life's dream, on her way to becoming an author, but who had lost her body and mind in such a way that she would never deliver the book. Instead of words, her life was pills.

At some point, I lost Alexander again, and he went like just another bad night, barely noted. It felt like he'd hardly been there at all, my on-and-off bicoastal long-distance boyfriend, something of a different life, a road not taken, and as if he knew, too, he just faded away, his concern becoming complacency. I feel so helpless. I feel like I don't know how to help you. I feel like I don't even know if you want to be helped. Something has happened and there's no us anymore. I don't think I exist for you.

He was right. I never even called him back. Instead I lined up the pill bottles neatly, like someone worshipping at an altar.

In the meantime, I gave up on trying the little things—the websites, the tips, the advice from friends. When one of my psychiatrists recommended a UCLA neuropsychiatrist with an interest in somatization, I agreed to see him, but I didn't tell anyone what I knew: that this would be my last doctor, that I was giving up and past him there would be no one else.

The best part would be getting rid of people and their expectations, all the concerned eyes that were on me all the time, from family to doctors to the few friends who still came around.

At some point, I even attempted to date someone, at the encouragement of my mother, who claimed she'd always seen me flourish most when a boyfriend was around, who saw it as a distracting element. But I was past that point. I had first met the young attorney Joseph for dinner at a local seafood restaurant and ordered what I did not want to eat at all: a whole fish. Nothing else on the menu had made any sense to me and somehow it had occurred to me that it might be nice to stare down at a beautiful thing like a fish. But when it came I spent the dinner in utter distress, picking at the tiny bones that felt like my own fragile body, chewing and swallowing mostly air.

Joseph thought I needed to have more fun and a week later took me to a concert at the Hollywood Bowl. As I was gasping for air in his stuffy car, I accidentally lost my ticket out the window on the highway. Miraculously we found it by a gutter on the side of the road after we desperately combed for what felt like an eternity, Joseph cursing through every minute. And then once there, I could barely handle the crowds and noise, and for the first time found myself reaching for the Ativan as a crutch in a time of recreation. I went to the bathroom and for some reason I felt it would make more sense if I snorted it—if I thought of it as a party drug, maybe somehow it would seem lighter. But as I diced it and rolled a dollar bill in the afternoon stall, trying to alchemize the situation from Ativan to coke and hell to party, it occurred to me I had fallen one tier lower. When I returned to Joseph in the pit, I tried to soothe myself with the thought that my days had to be numbered at this point.

Enjoy the very benevolence of a concept like suicide, I would tell myself, that final option that could restore control, that mercy every human has for themselves, something you can save, save for last. I tried desperately to bathe in the glory of abstract concepts, but especially that one, because it had not yet rejected me.

They say if a person has plans on how to carry out a suicide, you should take the threat seriously. I had plans. I had entire drafts folders in my Gmail devoted to it. Back then everything felt like a half-written email for which you never hit send, notes to myself piled on top of each other, fragments that were indecipherable, half thoughts of varying degrees of import, simple scribbles of words that were attempts at coherency, lists for my doctors, good-bye letters to my friends and family—in the drafts folder, they were all the same in their silence. But in spite of their casualness, I had plans, that was for sure.

The first was the most logical, that row of pill bottles on my desk, what was supposed to heal me could also easily do the flip duty of killing me. I had enough pills for several suicides, for the suicide of a small army. How easy, how obvious, how altogether tempting it was.

But there were other options, less close to home.

The walk into traffic. The theater of a mauled body. The death that would leave no chance of survival.

Or the self-immolation I once witnessed as a nine-year-old, at a protest at the Federal Building in Westwood when an

Iranian activist poured gasoline on his body and lit a match midspeech. It was almost a thing of beauty, I had thought. Something about its perfection, being devoured by flame like that, as if you were never even there.

The swallowing of various poisons and the torture of internal combustion, more a sort of punishment for living in the first place, somehow seemed fitting to me. I had become convinced that my shell was not something meant to contain me, that I would have to be released from this body sooner than I thought, and that perhaps it would have to be up to me. I also considered that it was because of bad fortune, because whatever was wrong with me was not to be uncovered in my lifetime, and I began to wonder how many others throughout history had given up because life had given them no choice. Knowledge could not be expected, but one didn't need to tolerate life without it either.

And then there was one last idea: illegal drugs. I constantly thought about the writer Lucy Grealy, who had died in 2002 of a heroin overdose after years of being ill and depressed. The perfectly, stealthily orchestrated death by a cousin of recreation, by the one drug I was always too scared to do that would no longer have the power of taboo against me. I could flood my system with it, and let it carry me away, to that place great poets also had gone, a sort of glittery sleep, I imagined, so heavy and so light at once.

I had so many plans, and I was so grateful to still have enough of a mind to make plans. But in the end it was not death, not boyfriends, not my parents, but that final doctor—who said he knew how to get me off the benzos so I could see what was wrong health-wise.

"But to be clear, I don't really think there is anything wrong with you physically," he'd add.

Eventually I stopped arguing. If he could get rid of the benzos, it would be enough of a start. Somehow I felt I could deal with all else if this one thing was handled.

He put me on a drug called Neurontin, which they gave pill-addicted and alcoholic patients in detox, and which also had mood-stabilizing properties (he felt it might be appropriate, as there was some chance I could be Bipolar II). I didn't argue, and since my suicidal ideation was so thick, I no longer felt fear with pills.

I took the drug and within weeks began feeling a calmness wash over me, sleep coming more and more easily as time went on.

By some miracle, I was improving.

*I don't know how to say this, or how to believe this really,* I wrote my best friend. *But I think I'm somehow coming out of it. I won't believe in it fully, but there is some evidence it is happening for me.*

I started to return to the manuscript. I kept on telling my editor in New York all about "chronic fatigue" as if I was buying time—plus, a part of me thought it could be true, that that and fibromyalgia were controversial diagnoses at the time, but I certainly had many of their symptoms—and that it would be done soon but that I needed that time. And it started to get done.

I even returned to the Rodeo Drive boutique to put in some limited hours and add structure in my day. Job, yoga, the occasional friend, the manuscript, the good pills, my good psychiatrist—life felt complete in a very precious way.

I never forgot I very nearly lost it all. But I also never imagined it could happen again. And again.

---

Ryan was the first person I openly talked to about it all. I told Ryan about reading of Grealy and her overdose, and how I was going to order heroin and how suicide would be the best, because I imagined it as a going into a deep sleep, what I had longed for for so many months becoming a forever sleep.

Ryan got it—he'd been a heroin addict for years. He'd lived all sorts of lives: junkie, art school student, Nashville pro-skater, Hare Krishna monk, tailor, and now fashion designer. He was friends with friends of mine and word was that he was very much looking for a girlfriend and wanted to settle down.

I met him one day after work at the Rodeo boutique, and I immediately fell in love with him. It was not just his beautiful brown skin and silver hair—it seemed impossible that he was a white boy, his body covered in jail-style tattoos, his oversized black hoodies, his weird drawl and skater style. I fell in love with him because he was the first person who immediately understood me. And he didn't drink, didn't go near pills—he only believed in cooking very healthy food, sleeping well, and living a grounded spiritual life. So much so that he did not do much in the material world. He was one half of a design duo with his ex-girlfriend and it was clearly falling apart.

I felt in some way that it was the *prasad*, the cooking that he prayed over just as if he were at temple, that helped bring me to wellness. The first time I was at his apartment, he served me *kitchari* that he had prepared. I took a bite into that delicious goodness and suddenly felt very light-headed. He lay down with me—only to stay by my side—and held my hand and seemed to understand I was healing.

"We're going to make you better," he whispered. "I'm going to be there for you."

And in the next few months he tagged forever onto that.

# 5

## NEW YORK

It wasn't the first time and it wouldn't be the last time that New York was the answer.

What was only months but felt like lifetimes ahead, I found myself in that old city of mine yet again in 2007, in front of bright lights, taking my professional author photo. I had known George, the photographer, for more than half my life. He had been my high school journalism teacher in Los Angeles, and shortly after I moved back to New York we had reconnected. He was an obvious choice for headshots.

He asked to see what I had with me for wardrobe. I opened my bag and out came dress after dress, silk organza, crêpe de chine, satin, Italian wool, all impeccably tailored black dresses. Dollar signs flapped their wings through George's studio. I waved them away. "Not what it looks like," I told him. "Just dating a fashion designer." I didn't tell him that Ryan was just months from walking away from it all in the fashion world, at the very end of that road of many roads for him, and was doomed to be unemployed through most of our relationship.

"Just!" George groaned, rolling his eyes.

I went to the bathroom to put on makeup, thinking of all the things I had to do that day—NYC could be so overwhelming when you didn't live there, but perhaps it was also that way when you did live there, only you got used to it—when suddenly my vision started to go dark.

I tiptoed back out, on eggshells with myself, and George brought me water, snacks, his hand, a hug, the perfect questions.

*I have been through so much*, I wanted to say, but the words were tangled in my throat. They were still with me, the summer and autumn spent suffering from serious anxiety, panic, depression, chronic fatigue, gastritis, carpal tunnel, God knows what else, all the shattered states in the nightmare nation of chronic insomnia. I recounted for him what he knew: how I had reached out and out and out, calling and asking for help, collecting anecdotes as cautionary tales—stories of hell, stories of hope, just stories to breathe with, to breathe through. With no use for the nighttime, I'd call people at the oddest hours and want words from them. *Please just stay on the phone; please tell me a story, any story*, George remembered me pleading with him late one night, or perhaps early one morning. In the worst times of my life, I could not imagine anything more powerful than my only business, my first love: stories.

To most people I didn't know well, I sold it as a sort of dark but not unglamorous period. After months of a strange sort of rehab culled straight from my California playbook—acupuncture, Chinese herbs, Ryan's beads and talismans, and my revived vegetarianism—I was happy. Happy to call back all the people I had held hostage on the phone for advice, the doctors I harassed on weekends, my parents who had been rattled by my depression since I arrived. Happy to call back my agent and editor and try to come up with new euphemisms for the old euphemisms of my condition. Happy to throw away pills. Happy to be in love.

"It's over now," George told me over and over, in his embrace that day. "You're safe from now on."

I had tried to inhabit that feeling—that reality even—for months by that point. Ryan and I had moved to Park Slope, and I prayed the upcoming summer was kind. Summer had never been my season, but so much of what was once mine was being quickly redefined.

My edits to the novel were done and it was now out of my hands and in purgatory before entering the world. I loved that phase: the middle of the road trip, someone else driving, seeing a world outside pass by, deftly escaping resignation to thoughts, assignment to words. All I could do now was wait: prepub reviews, then real reviews, then my first book tour. I had never thought past this point in my life. Everything had been about the becoming, whatever came before and up to the point of *author*.

There was of course the old constant, panic about my finances, especially now that my advance had gone to doctors and drugs in California—and the final trickle of my advance was doomed to coincide with my impending book tour. Ryan couldn't hold a single job either. So I applied for my first proper teaching job at a university in Long Island and was called back for an interview.

I was fine most days, but sometimes there would be old reminders of something just a bit unresolved, like that day at George's studio. Just weeks before the launch of my book, on the day of the teaching interview, I became consumed by nerves, in a way I hadn't been in months. But this time it was in more of a good way, unlike the summer before. This time, I had hopes. I assumed all the gods were on my team, since I hadn't been notified otherwise. I stopped by a Starbucks in Park Slope on my way to the interview, having taken on a pitifully ritualistic Frappuccino habit. I sat there, reviewing my teaching philosophy, which sounded miserably fake, and I agonized over how to make it sound less so.

When I stood up, there they were, the stars—not Park Slope literary luminaries, but again the hypoglycemia-diabetes-cancer-AIDS-godknowswhatIhave kind. I panicked. I didn't have much time before I missed the train to Long Island. I was worrying about this as my vision wiped out completely, immune to the aggressive sunshine beaming over the brownstone rooftops.

Fade in, and I was slumped on the street quite indelicately, with some young hippie chick asking me " . . . you okay you okay you okay?" Her eyes darted back and forth from my squinting eyes to my hair that was partially bleached white.

"You passed out," she said and pointed to the left. "Let's go to the hospital."

"Not a chance," I told her.

She protested, this girl I didn't even know. Eventually, I told her the truth.

She nodded sympathetically. "I don't have health insurance either. But still."

Somehow, she won and half carried me to the ER. Then, as if to reward me, she immediately disappeared. I let the nurse take my temperature and my blood pressure before I ran out the door. I called the university and told them I would be late.

I fell four more times that day, but I got the job, the only job that would put up with my book-tour schedule: an adjunct professorship teaching comp that paid a couple hundred a semester.

Back at home, I watched my unsteady hands at the keyboard

hit and miss over and over, not unlike the unsteady hands of my childhood, which had disturbed my mother so.

---

Ryan ended up being the perfect boyfriend for this period, just as he was the perfect person to usher me to health. He kept me sane by cooking his meticulous holy meals, cleaning around me, reading drafts of interview questions, and assuring me my recurrent insomnia was natural for a debut novelist.

Book tour, book party, reviews, all of it felt good at first, or good enough compared to so much bad in the seasons before. But it could only last so long. Two months after my fainting spell, I was sitting cross-legged on the dirty linoleum floor at the JFK Delta baggage claim, hugging my carry-on bag like it was a pillow and trying to sob subtly into my cell phone. I was crying about money, something I had a negative amount of, according to a robot at my bank. I had some change in my jacket, but it was not even enough to get a cookie from the concession stand in front of me and I was trembling with hunger.

I hadn't had money for weeks. My paperwork from the new university job had not gone through yet. My publisher had paid for some plane tickets and hotels, but all I had of my own was what my struggling boyfriend could spare. I had a million free fancy dresses to wear and a lot of good face to put on, but when it came to eating I was eyeing the prices on every menu and pretending cookies and chips were my foods of choice, that Subway was my adorably ironic passion, that the McDonald's breakfast menu was my kitschy little crush.

But the most disturbing part of overdrafting my bank account was that it resulted from a certain check, made out in the previous summer, that I had no memory of. It was a three-figure check, written out to . . . *my psychic*.

I called people, but I didn't want to ask for help. I wanted them to think of it as a humorous anecdote, but I didn't want them to think my life was that difficult. After all, certain friends who were not involved in publishing thought I was rich and famous now. Why burst that bubble?

In the end, I borrowed money from a friend of Ryan's and took that walk of shame to a yellow cab, when I knew there were buses and shuttles and subways and all sorts of only semi-impossible ways to get back to Brooklyn. I didn't have it in me for anything more complicated. The touring was exhausting me, beginning to break me down again, but I didn't want anyone to know.

Later, when my publicist found out, she was shocked. "Why didn't you call us?!"

I gave her some glossed-over answer, but I wanted to say *I don't know who to call, when to call, why to call.* I was learning everything over again. I was new to the universe, not just to being a novelist. I suddenly didn't know what the hell I was doing.

I didn't mention that with all the plane travel, the interviews, the readings, the parties, I was starting to wonder how much more of this I could take. Fevers were coming back, insomnia, an occasional touch of panic, plus all sorts of weird aches and pains that I would try to ignore the moment they popped up. I didn't have it in me to go down those roads again, and whatever was unsolved from before would have to stay that way for good. My life was not showing signs of slowing down, and there was also the fact that I could not afford anything extra to deal with, much less health care.

Weeks later, on a day off from teaching, I discovered that gold was at its peak value. I sold what was left of a couple of family heirlooms my grandmother had given me to an old Iranian man in the Diamond District, who listened to a fraction of my story, gave me a decent deal, and told me, "My boy in medical university; my girl, married and with baby. Your fault for being a starver of an artist, daughter."

---

About midway into the winter the "offness" confirmed itself. Ryan and I had begun drinking and smoking again. Those habits were coupled with extreme poverty—weeks where my adjunct salary and his Citigroup temping meant $10 between us for groceries. We'd decided to move out of the tiny Park Slope apartment where we shared a full bed in a room that barely had room for more. We moved farther out in Brooklyn, to Kensington, which was not that far from Coney Island, but many Manhattans and Brooklyns away from anyone we knew at the time. We delighted in the first floor of an odd Victorian we inhabited and made up stories about the doctor who owned the building and lived upstairs. We tried to make the space beautiful with what little we had, but by February I knew we were both in trouble.

We couldn't afford our life in New York. We were both working too hard for little money. We were starting to grow sick of each other. The space was starting to show its cracks, first roaches, then mice, then the smell of mold everywhere. And the certain sickness I had been feeling again began to assert itself.

I focused on us quitting smoking and drinking as a way to boost health. But I began keeping a diary of my sleep schedule and it started to show its patterns: instead of the old abrupt insomnia, now the hours whittled themselves away sleekly, my bedtime growing slimmer and slimmer. The cycles of depression and anxiety were creeping in again. I finally confessed it to Ryan.

"Well, what do you think it is? You got over it last time. I was part of that."

I didn't know what to tell him, but we both knew I couldn't go down that path again. I called up my UCLA neuropsychiatrist and asked for Neurontin again, which he seemed fine with supplying, not knowing I had ever stopped. The truth was I went on and off these meds with little faith in them—away just long enough to know addiction was not an issue, and back just long enough to know they could work again. There was little logic in my patterns—but it was not a problem, I told myself, as long as I did not obsess over them again.

I found a local acupuncturist and started going there religiously. She saw a lot wrong with my system, and at one point referred to an illness inside me that like a dragon sometimes slept and sometimes awoke.

"But what is it?" I asked.

She shrugged. "Does it need a name?"

I wondered about that. A part of me felt like it absolutely did need a name, that that had been a missing element of my life now for years—this odd sense that something no one had found was very off, that it might forever go unannounced. Another part of me thought I understood her—what would be the use? More doctors, more pills, more money, more roller coastering through the medical system? When did knowledge actually help in this area? I really didn't know what the right path was, but I decided to follow her alternative one.

I got a bit better, then I got much worse. By the late spring of 2008, just as some of my psychiatric issues got under control, I started to get fevers, high ones. At the time I was full of excitement because I'd just been awarded a proper teaching position a state away at Bucknell University, a visiting assistant professorship that would pay three times the most I've ever made, plus health insurance for both Ryan and me and our own big house in the middle of the woods. This excitement made it such that

I didn't notice at first that I was dropping weight and running high fevers. It took Ryan forcing a thermometer on me, noticing my face had been very flushed for days, that I was talking rapidly, that I had been sweating through the sheets.

The fever turned out to be alarmingly high, nearly 104, and it did not go away for many days, even with multiple fever reducers. We didn't yet have health insurance of any kind, so we went to a small clinic nearby with modest fees and they referred us directly to the ER. I tried to convince them I couldn't go, mentioning I would have proper health insurance in a few months and could look into things then. But the nurse was adamant that I had a serious problem and it needed to be addressed *now*. I told her I had no way of paying for the ER, and she lowered her voice and told me to go without ID and they couldn't refuse to see me.

The ER it was. They ran blood tests and some things were off—most notably my thyroid. The doctor told me I should see my GP and my endocrinologist but I told him I had neither, I didn't even have insurance, and that I'd be best off just ignoring it.

He told me I could not do that, that there were indications of thyroiditis and something autoimmune related, and that they'd have to keep an eye on it. I'd have to come back to the ER.

I've never quite been able to recall it with any clarity, but I have a hazy recollection of someone in that hospital saying something to me about Lyme. Some days when I think back to this half memory, I think I must have made it up—that I want it to have come up then, another opportunity for help, another light in the path—but I still can't dig up enough to tell me one way or another. There was perhaps a thought of it—in fact, there was reason to suspect that what looked like autoimmune thyroiditis then was Lyme. But this would not be pursued until years later, when my Lyme doctor found out that Lyme tended to attack my hormones and blood sugar the most—that the endocrinological was my most vulnerable area.

The next few weeks were a blur of battling fevers from bed, daydreaming of our new life in Pennsylvania, fielding interview requests from my phone, deliriously responding to emails, and eating whatever ambitious nutrient-rich meals Ryan put in front of me.

"It's weird that there is not a real name for what is wrong with you and yet something is," he said at one point, watching me eat his soup.

"I've always known something was off."

"Is this how you felt that summer?" He meant that awful summer before we met.

I told him I wasn't sure, and I'm truly not. I was a drug addict then. A huge insomniac. I was depressed and anxious. I always thought there was something physical at the root of it, but no one had found it, so I was resigned to thinking it had been created in the mind. I was okay with it even.

I was, after all, someone who created in the mind.

I slowly inched my way back to health the next few months. Ryan and I packed up the apartment in preparation for our new life, one that luckily would only be three hours away from this old life that we were not altogether done with. We hadn't mastered New York, not even slightly. But opportunity had come. For Ryan it meant he could go back to school and get closer to a college degree—college classes would be free for him since I was a professor. In the house we were renting he'd have large studio space for his art. He promised he'd get a job, but he also promised he'd help around the house while I got used to teaching full-time, with a real salary, a real office, actual colleagues I went to meetings with and sat on committees alongside. Things were happening for us.

The day we packed the U-Haul, my fever spiked suddenly to 104. Ryan begged me to stop by the hospital again but I refused.

"It comes and goes, this thing I have," I insisted. "I have to live with it."

"But what is it? We need to know what it is."

I thought of the acupuncturist and closed my eyes as if to will the fever away magically, and I repeated her words: "Does it need a name?"

# ON APPEARANCES

It's February 2017, and it's the memory of Ryan and his *prasad* that somehow brings me back to myself in a hospital bed in DC's Howard University Hospital. It's months after this book was due, just months before it was supposed to come out, but here I was sick again. I had attended the AWP (Association of Writers and Writing Programs) Conference in DC and taken on a lot—my panels and readings, plus filling in for others. I had been well for some months prior, or well enough—after the car accident of the year before and its subsequent intense Lyme relapse, I'd worked hard to take care of myself. I had weaned off the high-dose antibiotics (what they call in Lyme communities a "chemo-load") in the months before the election, but after the election I went back on them. All my identifiers felt at intense risk, and my stress was sky-high; I needed the protection. But at the conference in DC, I'd only been back on the antibiotics a few weeks and their effects had yet to take hold.

I had collapsed in the lobby of my hotel, a Marriott not far from the White House, and the hotel had assumed I was just drunk or on drugs. "Late-stage Lyme," I tried to blurt out, trying but failing to let them know I just needed fluids and oxygen, that I was likely simply worn down. No one who knew my illness was with me, and I didn't want to bother calling anyone. An ambulance arrived, and a black paramedic tried to take my vitals as I convulsed, foaming at the mouth, going in and out of coherency. When she finally got me to make some sense, I immediately expressed a worry: *Please don't take me somewhere where they are racists.* She looked appalled, and I realized she thought I was white. *You're going to Howard*, she informed me. I nodded, remembering that it was a black school, and thinking of my own black Muslim neighborhood I felt some hope.

There in the hospital I thought of Ryan again, the last time I felt I had a reliable "emergency contact." Ryan was always the caretaker, the nurturer, the man who had insisted I put his name on all these documents. For years after, I used to think of relationships in these terms: who you could put down as an emergency contact. The only danger I felt in my life, in those years after the worst of it, was the lack of an emergency contact. In the many hospital rooms I'd been in over the years, I never felt more alone than when I'd pause at those words and finally succumb to putting my mother and father, even though they were thousands of miles away.

At Howard, I had my first good experience with an ER doctor, an older man who knew Lyme well and who nodded knowingly when I repeated my concerns about race and ethnicity. He talked about being a black Muslim as he gave me fluids, as he gave me oxygen. He was worried about how high my heart rate was, but he understood when I refused the Ativan in my IV. *Please don't let that happen to me again—that will actually kill me.*

He nodded. He'd seen people like me before.

It was a godsend when someone would understand me in these moments. When someone would get that I'm a former addict. That I don't look like what you might expect. That I'm a brown Middle Eastern woman, though every time illness turns me white—thin and pale to the point where everyone congratulates me at my sickest as I transform to a white woman in appearance. I'd even chemically straightened my usually curly, frizzy hair to a stick-straight silky whiteness that winter, because it was causing too much joint pain in my hands to wash my hair. Every part of me in illness became the white woman of their dreams.

In the hospital bed in DC, as I went in and out of sleep, Ryan would come to me with his bowl full of prayed-upon nourishment, this white boy who knew something close to my part of the world, who had worshipped and sometimes even "passed" in the brown East, who could put this world into its context and dare to heal. And yet a part of me also remembered a certain disgust at his surface, his beads and wraps and pendants while he was just a white boy from Tennessee. Cultural appropriator? Cult devotee who aspired to be cult leader? Con artist? Cultureless desperado? I must have known something was wrong earlier and yet he was all I had.

Occasionally, I'd be woken by the ER doctor at Howard. He made me promise not to return to New York for some days. I said that would be impossible. He made me promise not to go back to work in New York for some days. I told him that would be impossible as well—what else did I do but work there? He made me promise to get checked out there by a doctor. I said I'd try. For years I had not been able to find a local doctor who would take me when he or she knew Lyme was on the table; even some infectious disease specialists wanted nothing to do with it. But I knew at that point I only had half a year of health insurance left and that I'd have to act fast. This new administration seemed intent on taking away health care, even the somewhat broken

Obamacare I had been on. While Lyme was not ever really covered by insurance, emergency complications like this very hospital visit were. I was someone who knew my life meant ER visits from time to time. The luxury of having university affiliation after all those years when I could not work, the ability to work and thus have health insurance, it meant the world to me. And I was at the end of that golden period—who knew what came next.

I still don't know.

# 6

## PENNSYLVANIA

I always wanted to write a screenplay based on the life Ryan and I lived in Lewisburg, Pennsylvania—population four thousand—based largely on how I imagined our neighbors saw us: an artist couple, fresh out of Brooklyn, takes on Pennsyltucky! There we were, within the first few weeks buying a navy 1989 Crown Vic in cash. There we were, both in often-shredded, ratty all-black clothes, art-nerd glasses, and full of prominent tattoos. Even with some of his religious adornments, Ryan was definitely a guy you'd imagine spent more time in American countercultures than anywhere else. And with him I reverted to my old suburban punk adolescence. The two of us looked like recovering Goths, retired punk rockers, misfit weekenders who owned an obscure art gallery back in the city. We did not look like we belonged there.

There was the conservative white-bread Union County hamlet in which we rented a large house from a professor on sabbatical, and there was also the Bucknell campus—also small, also conservative, also white-bread. We were as much a spectacle on campus as we were at home. In some ways this made us the coolest kids on campus, never mind we weren't kids. The faculty seemed to love that about us, my department chair touched by the fact I'd barely taught, colleagues astounded that Ryan was a high school dropout, and the students so curious about the life we'd often escape to in New York City over the weekends.

It was definitely odd to be one of the only Middle Easterners not just on the campus but in the entire town, and at times this weighed on me. No place more than Pennsylvania did I hear shock over my name; nowhere did I feel more out of place for not just my style but who I was. Ryan balanced us out a bit, but he wasn't the normal white boy either. I began to wonder at times if we were partially there to be a spectacle for a world that didn't get much outside exposure, even though it was three hours from several major cities.

But Ryan started to fall in love with small-town country life, not to mention the university, where he got to take classes and dreamed of getting a college degree. In the basement he built a sort of art studio, while I worked on the third floor, in a tiny writer's den the professor who lived there had built. It had the feel of a treehouse, a nest, and I was more productive there than I ever had been. Plus, we had money, my very real income, and health insurance, the ability to actually take care of ourselves. At the grocery stores, we were like kids in a theme park, and we reveled in the adult feeling of being able to buy whatever we wanted. We built a real domestic life for ourselves for the first time.

Much of our time revolved around me being at work and Ryan being at home, cooking and cleaning. I knew he had to get a job, but he was suddenly so dedicated to his studies. He was also dedicated to worship and had set up an altar in the home, and he soon realized there was an ashram in Port Jervis, not too far from us.

The Port Jervis ashram was along the Delaware River and we made one visit there. Our room had separate bunks—if you weren't married, you couldn't sleep in the same bed—and all the food came from their crops and many farm animals. We took long walks at night through firefly-lit forests and at dawn we woke up to see deer run across the many meadows. Years later I would remember that scene and the conversations the ashram workers had with us about ticks and how we shrugged it off. I felt safe at Port Jervis—I credited Ryan with healing me from my mysterious insomnia-driven, pill-filled suicidal summer. I felt grateful he had another outlet for worship somewhat nearby.

Pennsylvania was where we discovered nature as a couple, us two city kids. We had both become obsessed with an inn Ryan spotted in a fashion magazine, called Hotel Fauchère, that had been built in the nineteenth century and housed everyone from Teddy Roosevelt to Mae West. The town of Milford was also at the neck of the Delaware Water Gap National Park, which boasted seventy thousand acres. We didn't know much about hiking, but we loved long afternoons spent winding through woods, punctuated by networks of waterfalls and dreams, Ryan chanting mantras to Krishna, and me dreaming of stories. TICK CHECKS ARE ADVISED ☺, cheerful signs around the community would remind us, but I can't remember us ever stopping to do them. We weren't looking for problems. For a good while we felt very lucky.

*Our country home is our main home*, we kept saying. We loved playing house, playing adult, playing professional. It was hard to remember we weren't married, that we weren't planning for kids. Everything seemed steady and solid—and then one morning it all changed, just like that.

It had been a normal midweek morning in February—I

had an afternoon class to teach, and I had made our smoothies (Ryan liked to be the main cook, though I tried to do my part with breakfast). Ryan had been staying up late that week and had not gone to bed, instead moving a spare mattress that we'd had out for some visiting NYC guests into the living room and watching marathon movies all night. I'd find him in front of the television in his clothes each morning, and I began to wonder if he was okay. He had not yet enrolled for his second semester of classes, which I found odd, and that morning I gave him the same reminder I had given him nearly every morning since we'd moved to Pennsylvania: *Maybe you could take some time and research some jobs?* Ryan had thought of all sorts of things, from spiritual teaching to working as a tailor to construction to setting up his own art store or fashion design consultancy, but nothing had panned out and we both knew there were not too many job opportunities in the area. I had even encouraged him to find work in NYC and commute, and he'd occasionally go for an interview but never got called back. Ryan was a character, not an employee, we always joked. Everyone loved meeting him, but they just could not imagine hiring him.

That morning, as I took my smoothie upstairs and gave him his usual job reminder, I did not expect to suddenly hear a big boom, which turned out to be him jumping to his feet, I later realized. I did not expect the booming to follow me up the stairs and nearly tackle me at my own door, the green shake bouncing out of my hands and exploding all over my bookcase. "Hey, what are you—" I screamed, at first just screaming out of shock. But upon seeing his face—he looked like a complete stranger—I began to scream out of deep fear. His face was red and damp, hot with rage, his eyes flashing and animallike. He was saying something, words roaring and frothing out in a way I could not understand. He began throwing my books off their shelves, knocking my lamp, throwing my papers everywhere, and finally pushing me against the walls of that little study, my nest. I could hear myself sound unlike me too, whimpering, pleading, begging him to stop, negotiating even, asking why this was happening—not unlike how I used to as a child with a raging father.

I could think of little else to do but to call the police. When they arrived, they said they had to take Ryan into custody and I begged them not to.

"Ma'am, have you looked in the mirror? You are covered in bruises. We take domestic altercations very seriously here."

I nearly rolled my eyes—in all the chaos the one thing I did not consider was that I was hurt at all. But after they left with Ryan I dared to look in the bathroom mirror, and there it was: a generous cluster of bruises all around my shoulders and neck and arms, and even a group of them that made it look as if he'd tried to choke me. But had he? It took me months to begin to remember what had happened.

His screams, altogether animal.

Tripping over his limbs, or my limbs, hard to know which.

The books falling everywhere.

His hands on me, all over me, suddenly not feeling like his hands but some other man's.

My body in pain, at many spots at once, in a chaotic alarm.

My own screams, also animal, begging.

His tears, my tears.

The sound of things breaking over and over, like an earthquake had hit.

Everything I had known as normal coming apart at the seams, in slow motion, all before me.

Until: silence.

Police cars.

So much explaining to do.

His last look at me, while in handcuffs.

The police car driving off and me in that rubble, my body still pulsing, the horrible cavernous silence that swallowed me for months.

It was the end of Ryan and me. He moved back to Brooklyn and although he apologized, we both knew things would never be the same, no open door left after all that. I don't think either of us ever quite understood what happened that morning. He had a psychotic break, that much was clear, but where did it come from? Was that even a question to ask? My assumption was that he must have been depressed out in the country and never thought to really deal with it. There was strangely not much more to it, and many years down the road, when we were back in touch, we never figured it out, neither of us. At one point, all those years later, we joked about it having been Lyme psychosis, and we both laughed, and then we were just quiet for a long time. We never got resolution and we never thought it would be useful to try.

---

After the incident, my energies were spent on trying to look ahead, to focus on something positive. I could not get depressed again. I had a book tour to Italy coming up—a big deal for me, as Italy had been my sole foreign sale for my first novel, and they were flying me first class and had lined up all sorts of fancy press, photo shoots, and talk-show appearances. I clung to this escape, looking forward to it with all my might, as if my life depended on it.

In Rome, I walked the overwhelming streets, filled with oversized historical landmarks that all looked familiar, and was nearly bulldozed by the hundreds of motorcyclists blazing down every winding road, and felt unanchored. With every book tour stop, Ryan had been the person I'd write to for pages and pages on email, documenting everything. I realized that for years now, every stage of my life had been calibrated by romantic relationships—including the measurements of health and wellness. Being alone suddenly, at this point in my life, made me feel especially unanchored. Who else could begin to understand me but someone like Ryan whose emotional intelligence came from a near-mystical source? As much as I didn't want to admit to that dependence, it felt like I had lost a soul mate and needed a placeholder.

On this trip I found myself writing one man in the end: Jacob, the closest friend I had in my university town. Jacob had been hired when I was hired—I'd noticed him at orientation, a handsome cross between Johnny Depp and Jude Law, the sort of guy who was too handsome to me to be even an object of attraction. He was an analytic philosopher, and it turned out that he was also the best friend of my most disastrous fling in Baltimore in grad school, that Connecticut alcoholic named Jim who wrote the most brilliant prose of all of us. Jim had written me when I first got the job to tell me his best friend was also hired at Bucknell, and Jacob and I had slowly become friends. Oddly enough, Ryan had always thought there was something between us and had been certain that Jacob had a thing for me. *He's always eyeing you.* I couldn't imagine it, as Jacob was probably the most traditionally attractive man I had ever met, and plus I always felt so inadequate around him. He was wrapped up in Plato and Kant and metaphysics and the philosophy of perception, and I couldn't even say what any of that was.

When Ryan moved out, Jacob and I became much closer. We'd spend winter evenings at the bar, sipping at local stouts on tap and taking smoking breaks together. He helped me get rid of Ryan's stuff, the junk he had refused to dispose of, the basement art studio revealing little art but tons of litter. Jacob always reminded me I deserved better, that I was wonderful, and that I would make it through all this.

"I've never not had someone to write to when traveling," I told him the day before I flew out. "That's what I do—I write when away."

"Okay, well, you can write to me," he said. "I'll be waiting for your dispatches."

He was so genteel, so elegant, so cultured, the kind of man who could spin emails as *dispatches*. I ended up spending much time in my hotel room weaving tales of my lunch at a Tuscan countess's villa, an evening in Florence guided by an extremely nervous Welsh poet, the daily hair and makeup in Rome, how the Italians thought of me as too schoolteacherish, one makeup artist bemoaning all my potential and proceeding to paint me into a centerfold version of myself for a talk show.

"You are beautiful and everyone knows it," Jacob wrote me. I didn't feel this way but it felt good to hear.

It took only a few weeks back in Lewisburg for me to have the courage, after a big group drinking night with all our other professor friends, to take to email, my favorite medium (never mind it was the coward's too), with a question for him:

> Jacob, I'm sorry for asking, and I encourage you to just pretend you never saw this, but I was wondering if you also think there might not be something going on with us? Do you know what I mean?

He had a class early that next morning but wrote to me just minutes before going in, Yes, it has occurred to me too and I thought to ask you as well—I have to teach now but I promise more soon—JP.

And there was more soon. So much more.

---

When Dr. E, the infectious disease doctor, asks me about my history that very first time, I admit to the fact that I remember the years in terms of relationships, that I count the men on my hands like they were truly parts of my body, all in the end that I was made up of: the Alexander era, the Ryan era, the Jacob era. Every phase of my health seemed to have had a partner attached to it, I admitted. *Were you ever alone during those times?* he asks me, and I tell him there were those bad periods when I was with my parents, when I'd moved back home, but generally I was in the company of boyfriends. *What did your boyfriend at the time think?* he asks more than once when I mention a particularly bad spell. They serve as echoes of my memory, as witnesses, as invisible testimony. The complete ignorance of the body on the part of Cameron, all the times Alexander insisted I was fine, the constant caretaking of Ryan, the varying degrees of concern from Jacob, who maybe saw the worst of it.

In my head I was always terribly alone, alone in the harsh shell of the body whose states were impossible to translate to others, but for most of it, there was someone around. Someone who held the teacups, who brought the thermometer, who pulled the blanket over me, who drove me to the doctor's, who held my hand for the needle at the phlebotomist. They were always there, those lovers of mine, guardians of sorts in each their own way, who allowed me to think—on some level—that they would be the way out. Just like changes of location, my boyfriends would be how I could untangle from my own hopeless interiority. Outside of me there were all sorts of possibility; it was the inside that was the problem.

---

Jacob felt like the most Adult relationship of my life, and it was through him that I felt I had the greatest sense of hope. I was renewed as a visiting assistant professor for the 2009–2010 school year at Bucknell, but it was perfect for us: my new boyfriend Jacob and I could build ourselves a real life.

We decided to move in together in the fall, even though I'd taken on another sublet in case. Jacob wanted me to move everything over into his new place, a big spacious modern house in the best part of town, so we could "practice," Jacob said to me, always with a certain look. I knew what he meant. He was very taken with the idea of us being together forever and he often told me if we weren't engaged by the next New Year, he'd be shocked.

While away from him for the first time that summer for a residency in Virginia, only a few months into our relationship, I slowly began my second novel—but thinking of him was a major distraction. I'd put on a full face of natural-looking makeup for our Skype sessions, which were often the highlight of my working day. I decided that maybe we should take a trip together—maybe that was what we were missing together over the summer. I had become concerned since Ryan's breakdown about not being attentive enough to my boyfriends, and I decided I had to take a more active role in keeping them happy. There was an anxiety to my interactions with Jacob that had everything to do with Ryan, but an additional layer of anxiety meant never letting him know I was still thinking of Ryan.

Ryan: the Bad Boyfriend; Jacob: the Good Boyfriend. That was how the story was spun, that was our shared story. Gone were the good memories of Ryan, how he had saved me from that awful LA summer, how he'd been at my side through so much including a book tour, how he'd moved with me to rural

PA. The bad ending had taken over everything. Occasionally Ryan would still call just to check in and I rarely answered, not because I didn't want to—a part of me wondered what his life was like now—but because Jacob had more than once forbade me to do it.

So my thoughts that summer went to a couples' vacation, and Jacob was elated. We settled on a place we had never been: Tulum in the Mayan Riviera of Mexico. Everyone around our age was obsessed with it—Marfa and Tulum were the rage those years—and you couldn't read about a stylish young couple of that era without hearing about them cave-diving and doing yoga on the beach in perfect Tulum. At the time I shopped for bikinis during my writing breaks at the residency, which felt like a good distraction from my writer's block, and Jacob revealed himself to be quite the traveler and planner, booking us flights and cabanas and all sorts of activities in Mexico. I began to feel like this was the start of my real adult life. That I finally had arrived.

Our trip was everything we imagined—beach time, perfect fish tacos, wildlife and nature, yoga and hikes, postcard sunsets and solitude—and we returned carefree, not a problem in the world, just like in the movies.

Mosquitos were an expected thing of course—both in Mexico and in Pennsylvania—but on our first evening back we realized how disproportionately they liked me. We counted bites and Jacob had five while I had eighty-three. It was then that he noticed that one of mine did not look like the others.

"A bull's-eye," he pointed out.

I can't excuse how Lyme-illiterate I still was at that point, but somehow I had never bothered to fully understand the concept of the rash that would present itself in the shape of red concentric circles, as in a bull's-eye. But Jacob had grown up in Connecticut so he was worried.

"It's not necessarily a tick bite, it's not Lyme for sure, but it's worth being cautious about," he said to me.

I asked him what that meant.

"ER."

I thought he was kidding. "ER for a bug bite?!"

He nodded gravely. "It's what people do. Trust me."

And I did just that. Jacob was the boyfriend I always trusted. I was surprised by his alarm, but with good university health insurance I could humor the concern. We went to the ER and the doctors agreed to run tests, finding the bull's-eye suspicious enough, reminding us that Pennsylvania was Lyme country after all.

I suddenly recalled how at my residency in Virginia everyone was talking ticks. I never saw a tick once on me during that time—though I was never conditioned to see them—but everyone around me had been very alarmed about it all. Plus during our hikes in the Pennsylvania woods—I'd taken Jacob to Hotel Fauchère, too, and he'd also fallen in love with the Delaware Water Gap—Jacob was very adamant about us doing tick checks, which Ryan and I had never even thought about.

Jacob was convinced I had Lyme. *I hope not, but I think it might be it*, he kept saying. *In any case, we caught it.* That evening the Lewisburg ER sent me home with some antibiotics while we awaited advanced test results. And then the results of the Western blot came a week later, while I was in Wyoming at my second residency: apparently I did have Lyme, though they did not think it came from that bite.

It's amazing to think of this moment, the first time I got a diagnosis and how I reacted to it: with very little alarm. How much harm could a bug bite cause? My whole life I'd been covered in mosquito bites and I didn't have the sort of family that made a fuss over that. By this point, three decades into life, I'd seen all sorts of adversity from war and revolution to abuse and rape and car accidents and terrorist attacks—what could a bug bite mean for me? It registered as something—perhaps some official entry as an East Coast person—but it definitely did not register as enough.

I remember asking few questions. I was told to go to an infectious disease specialist on my return from Wyoming, but to continue taking antibiotics, and I did just that, without any additional research. A week later I was due to go straight to my third residency—this time Yaddo, which seemed to come with a very dire Lyme disease warning itself, it not being uncommon for residents to go the ER weekly with tick bites, a writer warned me. And so I went, but it was there that I suddenly got sick, and then sicker.

This was clearly not just the little bug bite of minor consequence I thought it was. And East Coast types, from residents to Jacob himself, assured me of its awfulness. They told me it would devastate me for a while. They didn't tell me how long.

And so as bad as it got that summer, I accepted it with little struggle.

There came a point at Yaddo when I could barely walk, which the urgent care there said might be "Lyme arthritis." Jacob tried to get me to move from my room and studio in the top floor of the famed Yaddo mansion to something on the ground floor, but they did not seem very taken with my case. Occasionally a resident would help me up the stairs, and at one point, someone had to carry me. On the East Coast, it seemed no one was particularly shocked by it all—it was a fact of life, and I had just been unlucky.

By the time I got back to Pennsylvania I had been on antibiotics for months and seemed to be getting better. Still, I went to Dr. E, when I was back, and he ordered more tests. At first, he told me he doubted that I'd had Lyme that time around—he said he thought I had parvovirus, a disease dogs and kids get normally—and that the ER test for Lyme was not a good one. He dug into my past and we investigated all sorts of possibilities over many hours and many visits but there was little certainty to what we came up with. This apparently was also normal. At any rate, he said I'd taken enough antibiotics to be rid of it—whatever "it" was—and that I was okay and could go on with my life.

I think when he saw me taking too much of an interest in the diagnosis—or any interest at all, as at first I'd been consciously trying to ignore extreme alarm—he let it go and reminded me that if you go digging too much, eventually you find trouble. And anyway, he said it was *highly probable* I got Lyme in California as a child and I should be grateful I'd lived so well with it. *Just get on with your life*, he kept saying. *Write those books you write!*

I was not writing at all. Somewhere around midwinter, I began to find myself losing a grasp on sleep again, though this time there was no Ryan and his healing stews to bring me back to myself. Jacob seemed to blame my sudden disintegration on a reactivation of the summer's Lyme, but I thought differently. I thought it could have been a reoccurrence of whatever had been wrong with me on and off for a while—that haunted summer before I met Ryan, for example, when I forgot how to sleep and slowly lost my mental and physical health completely over months—whatever had never been figured out.

A part of me did not want to know, did not want to begin the cycle of doctors and hospitals and internet research and supplements again. I knew there was nothing to do but to ride it out, as painful as it was. And I tried my best to teach my classes, through no sleep, through terrible anxiety and depression, and eventually through panic attacks. Jacob tried to make plans for us for the summer, for spring break even, anything to cheer me up and make me feel better, but I was inconsolable as I descended yet again into illness.

I could not believe I was breaking down again.

But losing writing was the worst of it all somehow. I could not form sentences, my imagination seemed crippled, plot and characters seemed so abstract, theme an impossibility, all syntax and diction just puzzles that were unthinkable to piece together. Jacob drove me around the county to acupuncturists and healers and nutritionists and yoga classes, but nothing seemed to really work, nothing except perhaps time.

By spring 2010 I was a bit better and it was clear my time in Pennsylvania was over. I could be Jacob's partner at best, as my job could no longer be renewed. Jacob was understanding when I began applying for jobs around the country, and he said that even with long distance we'd always be together.

Late in the spring, when I thought all jobs had passed me by, I got a call for a permanent position in Santa Fe. Jacob was elated—as was I. We had both dreamed of the Southwest, talking of our love of deserts and planning trips there together. I had a few friends who had moved there who spoke of the Land of Enchantment as just that. Jacob immediately began to scheme on how he could move there the following year after for his sabbatical.

"And I can't help but think the weather will do you so much good," Jacob said. "I think for sick people, desert climates, especially high desert with such good air, does wonders."

I remember feeling a bit irked at that suggestion, that I was a sick person, though I knew it was true. It was what I was, who

I had become, but I still was years from being at peace with it. I had so many alienating identifiers that I had no room for this new one, I felt. "Who are you calling a sick person?" I remember smirking a bit, trying to lighten the term's load on me, though meaning it. I did not want that life, I did not want to be that person, and maybe a part of me knew I had no choice.

# 7

## SANTA FE AND LEIPZIG

Everyone always talked of Santa Fe as a "healing vortex," something I heard long before I thought I'd needed healing. My first friend in Santa Fe was a shy girl named Haley who within weeks of knowing me told me she had discussed me with her healer, and if I paid $300 for the first session, her healer could get me on the right track. I was shocked Haley thought I was on the wrong track—which track after all was the wrong one? Was all the hardship there to demonstrate how unlucky I was, or was I actually a very lucky person given that I'd survived so much? I neither contacted the healer nor pursued a friendship with Haley.

Instead, Santa Fe was a sort of luxury desert Disneyland for me—even the air felt expensive. Every sunset felt like a special occasion, there were scenic hikes all around, the smell of sage and roasting red chile pervaded, and kiva fireplaces were ubiquitous. The city felt in a constant state of festivity. I had splurged on a fancy home to rent—my university salary was substantial, a permanent position, and I was freelancing more than ever, my income never higher than that year—plus a little went a long way in Santa Fe. The house was a big modern take on the old adobes, about half a mile from the Plaza, on a private gated alley off Artist Road. I had a bedroom balcony, a beautiful patio, radiant heat, brick floors, diamond plaster walls, and a gold Saturn I bought with Jacob's help parked in my own garage. It felt like the highest incarnation of adulthood I could imagine—to me, this was how an adult who had made it would live.

Everything seemed better to me out there—or nearly everything. There was one small problem that didn't go away, and that was the altitude. I was warned about the high elevation, given that Santa Fe was over seven thousand feet high, and was told that I'd have to take some precautions, like chlorophyll drops, portable oxygen, staying hydrated, and getting lots of rest. It hit me hard in the beginning—I ended up tipping into a display of cowboy hats at a boutique the first time I got winded there—and I expected it would go away shortly.

It never quite did.

Something felt wrong, but in a different way than I'd experienced in the past. Here I was in paradise and now I somehow could not handle the elevation. Could someone never adjust? I wondered. No one seemed to know. They did remark that Santa Fe either took you in or spit you out—everyone focused on the spiritual aspects of life there.

This sickness manifested itself in many ways for me—I often felt exhausted. I often felt faint. I often had headaches. I often had no appetite. I often could not sleep. Worst of all were the nightmares. I began experiencing sleep paralysis at times.

The possibility that this was not altitude sickness and that it was something else was present in my mind. I felt acutely the insecurity of the unresolved illness of that Los Angeles summer in 2006, not to mention the trials of Lyme from Pennsylvania and its vague finale. But I'd shove it out of focus—*no new problems* was my mantra. Why go digging for something else when altitude sickness was a known hazard in the region? I tried to stay focused on the knowns.

But Jacob, who'd visit every few weeks from Lewisburg, was especially insistent I go to a doctor. He hadn't forgotten my months of struggles with Lyme last year, even as I tried to push them out of my mind.

Since I had made a group of well-connected friends in town through the network of fellow writers, I asked for a recommendation for a doctor. The arts college I was teaching at provided me with excellent health insurance, so I could afford to investigate further. The doctor everyone recommended was actually a nurse practitioner who was an expert in women's health. On top of that, by strange coincidence, she also happened to be Iranian. Firoozeh was a woman in her late fifties, quick-witted, sarcastic, sharp, and even a bit harsh, and she was determined to find out what was wrong with me.

"Why do you think there is definitely something wrong?" I asked. "Couldn't it just be altitude?"

She looked at me like I was crazy. "No," she replied. She was in particular fixated on my metabolism, the fact that even in my early thirties I could still eat an entire box of pasta for dinner every night and never gain weight. I saw it as something to be proud of, while she saw it as something to be concerned about. She tested me for all sorts of diseases of the metabolism, particularly ones that affected women, and everything came out fine except for one thing: my blood sugar. I took the five-hour glucose fasting test, and it revealed some abnormalities with my insulin and blood sugar, showing both insulin resistance and hypoglycemia. The test revealed that I was operating at seizure and coma risk. Firoozeh was surprised at the severity of the results, so they retested, and again I got the same disturbing results.

This led Firoozeh to think I had PCOS (polycystic ovary syndrome), or possibly something worse: a pancreatic or adrenal tumor. She decided to put me on a special diet to see how I'd do before additional testing. I was to go paleo, which was a real challenge out in Santa Fe—imagine no grains or sugar in the land of chips and salsas and margaritas and green chile croissants. But I abided, and indeed within weeks I felt better. She eventually decided I had something called "Diabetes 1.5."

Meanwhile my new job was turning out to be stressful. The arts college that had hired me was bought by a for-profit institution, which I had little understanding of in those first months. But soon I realized it was indeed run like a ruthless business: students were referred to as "the product" in an employee handbook, and Human Resources—there was an HR—referred to my department chair as my "business manager." There were all sorts of mishandling and negligence of students, but I had to sign an NDA.

I started to see my time there as limited, which was disheartening, given my dream house, my big salary, my beautiful city, and the fact that I was finally showing health progress. Jacob urged me to put things in storage, so that we could scheme on ways to return—perhaps he'd even look for teaching jobs at St. John's or in Albuquerque—and I hunted for other opportunities in the meantime.

A fellowship for authors in Leipzig, Germany, popped up: the Picador Guest Professorship. It was a big honor and offered a substantial stipend and one could be gone for seven months. It included free housing, minor teaching responsibilities, and covered the airfare. Jacob was immediately elated because as a teenager he'd done a summer program in that very city. Leipzig was old GDR Germany—very much still an East German city—a college town, a classical music and opera town, and he was sure I'd enjoy the weirdness, the history, the culture. He was also excited because it was his sabbatical year, and that meant we could go together.

As he'd originally forecast in our first few months of courtship, Jacob and I now had plans to get married. We'd become engaged several months into my time in Santa Fe, his way of sealing our long-distance relationship, and I was more excited than I could have imagined, given that this was a fantasy I didn't even know I had. But a darkness tinged this enchantment: Jacob and I had started having bad fights that seemed linked to his drinking habits. I had sworn I wouldn't tolerate it and left for New York to see friends, who made me promise to leave him. But the feelings of isolation I had in Lewisburg didn't allow it—I went back to him, especially when he promised to quit drinking. He would quit off and on and it always dictated the tenor of our union.

He had promised to stay sober if we got married, but the summer before Germany, 2011, the bad fights were occurring with such frequency that I began to rethink things. "What if we could just remove the pressure of marriage?" I suggested. I had never been the type of girl to dream of myself as a bride. A wedding meant little to me, and I had tried to convince Jacob that we needn't do it legally, especially as the gay marriage debate was at a fever pitch in the national conversation, and I hated this inequality.

I remember Jacob eyeing the sapphire and diamond ring of his mother's on my hand—which I'd never liked, wearing the garish adornment reluctantly as a duty—and looking so wounded. But in the end he agreed, and we decided to simply tell people we'd postpone our marriage till after we came back from Europe. I held on to my wedding dress and the stationery for the invites. I continued to wear his ring and discreetly monitor his alcohol consumption. We had both taken to Zen Buddhism during my time in Santa Fe, and we even saw a Zen relationship counselor. It was his idea, but I followed through with it all the way. I started pushing us to do more group meditations, and I hoped we could continue this in Europe.

"It will be the best because we will be together," Jacob said, and in that I heard the echoes of Ryan, about to leave his New York life to go with me to the Pennsylvania woods. Men were always going along on my rides, and my rides often flung me places I'd never imagine, paths I'd never naturally take; at the same time, it was unclear if they were caretakers or protectors or additional stressors when life would hand me its trials, trials these men couldn't access as they were primarily trials of the body.

All I could tell myself was that this was the life I had dreamed of. On the outside, every element was in place.

---

We were to fly out to Germany from DC, so we went there for a few days before our departure to stay with friends. It was then that I started feeling an eerie throbbing on my head, coming from a large bump that was likely a pilar cyst. I'd grown up with pilar cysts all over my head, but this one felt hot and full of pressure. I wondered if a cyst was rupturing, which I'd experienced only once in Los Angeles—I was sitting at a dinner with Alexander, when a pungent white toothpaste substance began oozing out of a wound on my head. But this time it wasn't rupturing, it was just heating up and building pressure. I began to develop a fever as well, which was when Jacob insisted I go to urgent care.

Urgent care determined it was a scalp infection that needed immediate attention, and so we were suddenly at Sibley Memorial Hospital, with me grabbing on the railing of a hospital bed while an internist squeezed the pus out of my head with her hands, wearing goggles that were especially designed for exactly what happened—the pus bursting everywhere. There was the awful smell again, but this time the incision required stitches and antibiotics.

We mentioned we were due to fly out internationally in a couple days, and the doctor seemed concerned. She said as long as we tended to draining it once we landed in Germany, then it should be okay, but that it wasn't entirely advisable.

We risked it. By the time we landed at Berlin Tegel to take the train to Leipzig, I was feeling very feverish, and the stiches felt tight and hot. To top things off, Jacob and I were bickering. It was not an auspicious start to our European year together.

In fact, the first few weeks of the semester I spent going to the local hospital for all sorts of wound cleanings instead of teaching. The doctors there seemed intent on making plans to remove more of the cysts, just as they seemed determined to find out the cause of my abscess, which was supposed to be fairly rare. Between my high school German and Jacob's more advanced practice, we got the information that we needed. Meanwhile, I wrote Firoozeh back in Santa Fe of my health misadventures, and detective that she was, she was fascinated. But she couldn't quite put it together—insulin issues did not mix with scalp abscesses. What was wrong with me? That investigation would have to be delayed again. She, like Jacob, believed in our return to Santa Fe.

The next few months we again strived for all the normalcy we could possibly create. I tried to stick to a paleo diet—a challenge as the Christmas markets began to go up—and Jacob tried to avoid alcohol, which was its own challenge in the land of bier. He slipped more than I did, and it started to become a problem again. His lack of a purpose there reminded me a bit of Ryan's, and so it made sense that his drinking could be tied to his lack of work.

Then one night Jacob snapped, but it was not quite like Ryan's psychotic break. It was something slower, gentler, something we'd had coming for a while. We were doing our usual catching up with cable TV through illegal streaming one evening, when Jacob polished off his beer and began shouting at me without much reason. It was not the first time this had happened, and already twice our neighbors in the university guesthouse had complained, both times older men who did not expect to hear loud yelling in housing meant for the best of the academic community in town. I had been so embarrassed—between my abscess and the neighbor complaints, just a few months into my time in Germany, I felt so much was off and I dreamed of returning to America.

But that night when he snapped and we both walked away from the computer that was bleating *Breaking Bad*, something changed. I remember I did what I usually did when he raged—went to the bathroom, locked the door, and sat on the toilet seat as if waiting. It was something I'd done since childhood, my private place of contemplation, my only retreat when the men in my world—starting with my father—lost their heads. I caught myself thinking that if only I had made it to the bathroom after Ryan's psychotic break, maybe he would still be in my life. But this time in the bathroom I discovered how distraught I was—tearlessly, silently crying, as if this form of grief came from a deeper well, somewhere dark and cold in the gut, where the tears had long run out.

My eyes lingered on the nail scissors by the sink. For a moment, for the first time in my life, I understood what it was to want to cut oneself. I felt the urge, icy and crisp, the clean gash, the release, the sourness of the pain, the bitter sting, the draw of no going back. I immediately shocked myself—was I that mentally ill now, and so suddenly? Was this where our relationship had gone? Was this what I had amounted to, now in this weird winter in Germany, so far from everything I had known, in a small space with my supposed love—and my thoughts ending up in self-harm?

When I exited that bathroom I was a different person—or I promised myself I had to be a different person. Jacob and I were over—that I promised myself too. I determined I would slowly but surely rid myself of him (I was scared of his anger), even if that meant finishing off the year in Europe by myself. I vowed to take myself back.

By the time December of 2011 came along I saw every ritual as our last—visits to the farmers' market, a trip to Prague to see friends, a trip to Dresden to see a Christmas show. He wanted us to go see his mother back in the States for Christmas, but I had to be back to teach at a low-res MFA program in Connecticut over the holidays and I refused to merge it. When we packed for break, maybe part of him knew he was leaving for good—he had packed nearly every last belonging. I never asked him why and he never explained. Instead I focused on what was coming. In my mind I rehearsed it—it would be over the phone, and I took that moment in the bathroom with the self-harm fantasy as my justification for picking such a coward's way out—and I heard my voice firm and strong and his weak and cracking, and I told myself I would never speak to him again. I told myself it was not just normal but a good thing that I could not imagine my life past that point.

In the end, my voice was weaker than I had hoped, but

it happened just as I thought it would, days after Christmas, days before New Year's. I was house-sitting, actually plant-sitting, near my old Brooklyn neighborhood, Ditmas Park, for old friends who happened to be newlyweds. Something was off with me the entire trip and I attributed it to the fact that I was finally parting with the man I had nearly married—and parting this time for good. Any time I doubted it I tried to envision that moment in the bathroom again, the scissors, my wrist, the soundless bawling. But there was something else, too, and that I attributed to jet lag, drinking large amounts of NyQuil to sleep and DayQuil to wake up.

While in my beloved old city, I carried out none of my plans to see friends and instead stayed in a lot. The one thing I did do was take the subway back and forth to different movie theaters; I made sure I saw at least one new movie a day. I told myself this was because in Germany they did not have the new American movies and it would take ages for them to get there, but really I think I just wanted a dark place to be alone and away from my thoughts.

In New York, I felt relieved without Jacob. And when I got back to Europe that January, I thought things would be looking up. But I realized I was returning a very changed person. Upon arrival, something still felt off with me, something deeply wrong this time. I was depressed, I thought, that had to be it—who would not be depressed when their longest relationship was over? Certainly I still loved him. And now I was in this country alone, in the dead of their notoriously awful winter. Jacob had known more German than me, and he blended in there physically, too. I looked more like the Turks they resented so much. Leipzig, they said, was full of neo-Nazi activity and I had always been a bit paranoid in those streets. I told myself that it made sense, then, that I would feel strange, apprehensive, and trapped.

I was staying indoors as much as I could, rushing home from teaching, avoiding situations where I'd have to interact with anyone outside of my colleagues and students. The American opera singers across the street who had become my friends last semester called and invited me places, knowing I was alone, but I always had an excuse to avoid them.

I tried not to humor the thoughts that this was more than depression, but I still felt ill in ways I could not explain. I decided exercise would be the solution. So I took up running again, something I'd done with Jacob back in Pennsylvania during a particularly bad winter, when I worried all the darkness and coldness was going to get to me. I started to enjoy treadmill running back then, a little bit like an addict—that infamous runner's high indeed had hooked me. This time, I couldn't bring myself to join a German gym, so I decided to try running outside. I bought winter running gear, or what I thought was winter running gear—I avoided salespeople, just grabbed things and checked out—and I took to the modest-seeming loop across the street in Friedenspark.

The loop was full of runners—Germans loved to run I thought, and they seemed to enjoy winter running especially. That first day I ran as fast as I could—not like I used to, pacing myself carefully, checking my rates and speeds and reading up on marathon training. I ran without dogma, without theory, without a goal. I ran like someone running away from something, as if my demons were literally chasing me across the park. I ran crying, tears streaming down my face, and one time I caught myself yelling, a strange canine yelp that felt strangely good to release into that cold wind.

I began to run this way regularly, eventually daily. Because I was doing it so recklessly, I didn't get too alarmed when all the running began making me very sick. One night I was talking to Jacob over Skype in the US—we were trying to be good friends to each other, in spite of my original plan to never speak to him again—and I remember mentioning that my heartbeat would not go down after a run, that it seemed elevated for hours, even at rest, like I was still running. "That doesn't sound good, Peanut," he said, the old nickname always holding, even then. "I think you should look into this." I tried to shrug it off, but it felt good to hear his concern. He made me promise to check it out and reluctantly I agreed.

My gut instinct said something was wrong, but I needed it to be nothing. After the weeks I had spent upon arrival dealing with the abscess on my head, I did not want to go back to doctors there—and this time I'd be all alone, with my clumsy grasp of the language. Also, a part of me didn't want to know. I didn't have room for new problems. I had just a couple more months left of European life and then I could begin again, whatever that meant. Until then, I didn't want any setbacks.

But when my sleeplessness grew to the point where I was too exhausted to walk, much less run at all, I knew I had to do something. I asked a program assistant about therapy. He seemed immediately concerned and then quickly assured me Germans loved therapy—it was meds that they were more wary of. But of course, I craved meds, mainly for sleep, which had for so long been my problem. I took an English directory list he printed for me, and after much investigation I found a therapist who was also an MD, and apparently a leading researcher in anxiety in Germany. I wrote him a brief email and he responded immediately.

When I first met this doctor and his junior colleague, they were professional but a bit gruff, like so many people I met in

Leipzig. They had reviewed my notes and they thought I could use some talk therapy. They mentioned they understood how alienating Leipzig could be for an "outsider," and they apologized for the town. I lied and assured them I loved the town—it was just that I had some health issues (I mentioned the abscess) and my worrying was causing sleeplessness, because my fiancé and I had split and surely that had done something to me, and did they prescribe my old favorite, Neurontin? (I had done some research online and could see that they did, calling it by its generic name "gabapentin.") I needed meds, and there was no way talk therapy—especially with a language barrier!—was going to replace that need. For nearly a half hour they tried to convince me of natural solutions and holistic medications. As they rattled off the benefits of valerian and chamomile, I nodded and appeared engaged, but somehow by the end I still got a prescription for my medication. I assured them I'd only try it after the natural stuff, but I took it immediately.

For a brief time, the Neurontin once again helped. I didn't know what it was doing or even what it had ever quite done, but it was something familiar and something that at one point had not harmed me, so I felt comfortable with it. I felt optimistic. But then in the weeks to come I began getting dizzy. I rapidly lost weight, and I collapsed twice on the streets of Leipzig—I couldn't tell if it was fainting or a seizure. At the second incident, a neighbor told me to go to a doctor across the street who dealt with diabetics, without knowing about my blood sugar issue. In a daze, I went in and found Dr. Schreiber, a young doctor who spoke good English and said he could help me.

The next few months were spent ping-ponging between Dr. Schreiber, the diabetes doctor with the neighborhood private practice, and Dr. Mulder, an endocrinologist at the university my therapist recommended. (Guest professors didn't have the same access to health-care benefits as residents.) There were many blood tests, which luckily cost a lot less in Germany. The test results seemed off here and there, but there was no consensus. The doctors tossed around hypothyroid, insulinemia, repeated Firoozeh's PCOS suspicion, and echoed anxiety and depression. They tried to put me on thyroid medications, birth control, antidepressants. I was given a blood sugar monitor and was told to measure my blood sugar several times a day and keep a food diary. I did that, but none of it led to anything conclusive and definitive.

In between trying to be a guest professor, and making appearances in town at events and festivals, I had this other life: one of endless mind and body investigation. At times, I wondered if I could talk myself out of it and just save it for when I'd be back to the United States, where communication could at least be clearer. But every time I'd resolve to give up, something in me would tell me I couldn't make it. There was no figuring it out—it seemed to stump everyone.

Could my mind have caused this? At times, it seemed yes and at other times no. I had so many symptoms, every day felt like it brought something more, so it was hard to analyze. I started Skyping and emailing more with Firoozeh, who was perplexed by my rapid disintegration. It seemed like a puzzle she wanted to solve, and she was on my case more than I was, writing and calling daily. When I wanted to give up, she kept pushing. Somehow her presence in all this, even remote, was a comfort.

Finally one day she started asking me about environmental factors and pollution and I shrugged it all off, until she brought up something new.

"I can't help but think since it's Germany they may have a black mold problem there."

That rang a bell—apparently the university was very paranoid about it. When we first moved to humid Leipzig I still recall Jacob shaking his head at a clause in the paperwork, which said we must air out the apartment and always keep the bathroom window open, in case of mold from the shower.

I put her on hold when she said that and dashed into the bathroom to look. And indeed I found those few spots of black I was looking for—not quite like what I saw online, but something for sure.

That had to be it. Firoozeh thought it made perfect sense, and she was very frantic about it, sending link upon link and all-caps messages. It seemed I had to act fast, she kept urging me. I began sleeping at the home of the opera singers, where sleep was a challenge because it was a new environment.

When the university heard of it, they were alarmed and did their own investigation. In the end they denied the mold was the pernicious kind, but who knew how a compromised immune system could react to things—I had arrived with an abscess, after all, and had been ill off and on ever since.

My parents called frequently, and during one call around this time I just lost it—crying for several minutes straight, wordlessly, to my mother. She decided she was going to come and help me pack and get me out of there—at this point I only had a few weeks left of my time there, and my teaching was basically done. I tried to tell her it was okay, but we both seemed to know I was too weak to handle packing and leaving the continent alone. And of course, she had done it before with me, in Chicago, lifetimes ago.

By the time she arrived in spring of 2012, there I was and there I was not once again: I was that ghost, barely there but trying somehow to smile and make jokes and appear well. But every day around midday, when the worst of it would hit me, I'd become nearly catatonic with a mix of exhaustion, panic, and pain. (Years later my doctor called this "dysautonomic crashes," something I still deal with at times.) My sleep had been reduced to an hour or two a night, but I was too fearful to take the Klonopin that a friend of the opera singers had given to me. I had saved it to take once my mother was there, *in case*. I didn't know in case of what, but finally even my mother urged me to try it. And as I lay there trembling in the dark—the same poison medicine of the class of drug benzodiazepines that had caused such violent dependency in me five years before, in an era of similar desperation for sleep—I waited for its old familiar wave to wash over me and take me away, as if this was a little suicide. But I never felt it the way I used to—sleep overtook me before I could notice it, and then I woke up twelve hours later.

Instead of refreshed, I felt disoriented in a very scary way. I didn't know what happened—where was Jacob? Were my classes done? Why was my mother there? How was it time to pack? How were we going to get home? What had happened to me? How could I rewrite this whole mess? This went on for days. It felt like a part of my brain had become erased in that long sleep, the longest I'd had in months. Strange apparitions flooded my brain, burrowing in every crevice. I began to see things in the corner of my eye, flashes, abstractions, animals—I was constantly jumpy. I began to have weird theories as well, and the one that stuck the most was that I was sure I'd die of a heart attack or stroke if we took that transatlantic flight home. I could see it, I was possessed by visions of it. My mother tried to wave it off, but I was inconsolable. I imagined the flight turning back and then me getting on another flight and it also having to turn back, all because of my illness; I imagined dying on board the flight and my mother carrying my body to the baggage claim. I also convinced myself that I needed to increase the Neurontin, take more doses a day to steady myself—the withdrawals in between doses were getting me, I told myself. I started to find all food suspect and sleep seemed ever more troubling—I never wanted to wake like that again, with my self left behind in some other dimension, maybe never coming back.

Something had gone horribly wrong, I finally faced it. I thought of myself in that bathroom, that autumn evening, with Jacob bellowing outside it, thinking of the scissors at my wrist, and for the first time remembered what Jacob said when I came out, slowly but determinedly full of confidence in the future.

"Everything needs to fucking change," he was saying. "We are more fucked than we think." And he was right.

# ON PLACE

In telling this story, it occurred to me that it wasn't character or plot or even theme that was the ruling principle of its composition, but something far less likely: setting. I wondered why setting would feel so foreign to me, when location changes have been more than simple set switches for me. One could imagine the variations in physical location are what in some ways got me to illness—and Lyme disease—in the first place. It wasn't Iran, but then was it California, was it New York, was it Pennsylvania, was it . . . where? I would be destined never to find the bite on the location of my body, just as I'd be destined never to know the location I was in when bit by the tick. The question of *where* would be the most mysterious of all.

In recent years I've had many discussions with various doctors about Lyme and have seen how their opinions and protocols would shift. Usually they would land on this idea: *the truth is we know very little about this illness.* It would drive me crazy, as if it were all guesswork, and I've found that I've never completely abandoned doubting it. But there's something I've held on to, that I've heard from more than one doctor: the link between chronic illness and PTSD. If post-traumatic stress disorder can be defined as "a mental health condition that's triggered by a terrifying event—either experiencing it or witnessing it (symptoms may include flashbacks, nightmares and severe anxiety, as well as uncontrollable thoughts about the event)," according to the Mayo Clinic, then it makes sense that it can trigger a relapse or surfacing of some disability or illness. The disability or illness itself could cause the PTSD, and the experience of going for years undiagnosed and then misdiagnosed as many like myself do can cause considerable trauma, to put it mildly.

It is no coincidence then that doctors and patients and the entire Lyme community report—anecdotally, of course, as there is still a frustrating scarcity of good data on anything Lyme-related—that women suffer the most from Lyme. They tend to advance into chronic and late-stage forms of the illness most because often it's checked for last, as doctors often treat them as psychiatric cases first. The nebulous symptoms plus the fracturing of articulacy and cognitive fog can cause any Lyme patient to simply appear mentally ill and mentally ill only. This is why we hear that young women—again, anecdotally—are dying of Lyme the fastest. This is also why we hear that chronic illness is a woman's burden. Women simply aren't allowed to be physically sick until they are mentally sick, too, and then it is by some miracle or accident that the two can be separated for proper diagnosis. In the end, every Lyme patient has some psychiatric diagnosis, too, if anything because of the hell it takes getting to a diagnosis.

My PTSD was always tied to setting, with my family leaving Iran at war and revolution and coming to America as foreigner refugees on political asylum. There was never a home for me as a human in the world—which is why moving around was almost easy. There was never a home for me outside as there was never a home for me inside—my own body didn't feel like my own. There was never a context in which I got to know it, to be at peace with it. Only recently do I wonder if that has to do with being considered "foreign" or an "alien" or "outsider" or a host of any other less kind terms for us. How could I recognize myself if no one else could?

Los Angeles and New York proved to be homes of sorts after Iran—suburban Los Angeles where I was raised and grew up, and New York City where I came to adulthood and found my career. But even there I would lose myself at times, which made trying to pin down the body and its conditions all the more impossible.

My Lyme relapses almost always coincide with global turmoil. It was no wonder to me that I would often become sick after some external political stressor, like the Paris attacks, or the election of Donald Trump and its endless horrific aftermath of mainstream bigotry. When the Muslim ban became a constant on the news in 2017, when I found my home country rather unsurprisingly on the list of six countries that had been designated problems according to this administration, I immediately had an acute plunge in health. I've been back on antibiotics since and now can't imagine being off them, not as long as Iran is in the news like this, not as long as all the trauma of my childhood comes back to me in waves these days. It's a return to hearing "go back to your country" regularly, especially online, and it feels like the most familiar nightmare. When I feel myself getting sicker as a result of the news, a part of me panics—is this just psychological? Was it just PTSD all along? Were some of those early doctors right, the ones who just thought I was crazy? How could my body erupt in a chaos of spirochetes each time my mind and body suffers? How does that work? And yet we continue to find evidence of the mind and body connection.

It has taken many years to see my own shell, this very body, as a home of sorts. I can report that even now I struggle with this concept, that even as I type these words, something feels outside of myself.

I sometimes wonder if I would have been less sick if I had had a home.

# 8

## LOS ANGELES

Coming back home felt like something else altogether this time around. First of all, home was not home. My family had bought their first property since coming to the United States and we were no longer even in my hometown—instead of Pasadena, this was more affordable Glendale next door, and this was no longer the room I grew up in, but a spare bedroom that was a sort of compromise between my brother's and my rooms, a generic little space, with a twin bed and a sort of impersonal ambiance. The place was cleaner and more modern, but it was not comforting in the least. It felt like I had walked onto the wrong set.

I was in a state of horror from the first day I arrived—trying to organize the space in a way that could feel mine and then reorganizing it all over again, pacing as if looking for something and then forgetting what I was in pursuit of, imagining myself out of there before I could even commit to being there—even though I was back to having no imagination about my future. I told everyone I knew I had black mold poisoning, plus a strange dependence on medication, and I started finding myself reverting to that damaged person in sunglasses and hoodies from a few summers prior, the one who was haunted by sunlight and doomed to insomnia's endless nights.

I tried to remember all my recovery methods from 2006 and go through those motions again, which meant reestablishing contact with my psychiatrist at UCLA. He saw me that same week and seemed disturbed by my state, which he noted was similar to how he first saw me and possibly worse. He seemed exhausted by my insistence on physical problems, the poisoning and other potential ailments—and so I didn't mention Lyme, thinking of it without it ever hitting my lips. Other things seemed more likely to him, and yet at the same time none of them seemed a proper contender.

"I don't think we should stray away from the possibility that this could just be a psychological event," he kept saying. He had me replay my life in Germany over and over, to which he often said, "Well, that doesn't sound good, does it?"

"What doesn't?"

"All of it," he said, matter-of-factly. "It seems you were exposed to significant traumas in Germany."

I didn't know what to say to that. There had been good things—my classes, my students, excursions to nearby European cities, the Christmas markets, my decent salary, even the guesthouse held a sort of charming sweetness in my mind.

*A psychological event. Significant traumas.* The other thing I didn't mention was the night in the bathroom, the contemplation I had of cutting. Given his reaction already, I didn't see the point. I had been damaged, certainly, but now I was going to get better. What did those memories matter now?

He decided it would be prudent to wean me off Neurontin and put me on Klonopin. I was shocked when I heard this.

"Don't you have your old notes there?" I asked.

He looked almost amused. "Of course I do. Why?"

"Well, you got me off benzos—you weaned me off Klonopin by putting me on Neurontin. That was the whole point of it."

"But you weren't on Neurontin to get off Klonopin this time around. Neurontin has many uses. I'm not sure why you were on it."

Neither was I. We were both silent.

"I was a benzo addict," I said. "*You* put me on Neurontin. And now you want me back on benzos?"

"I want us to get a handle on your sleep," he said. "Which feels urgent, so we can figure out what else is going on. Klonopin is relatively harmless. And we can get you off it again. I need you to trust me, as you did last time. We did it, remember?"

I did remember, but more than the memory of recovery, I remembered the memory of addiction. I put the prescription he'd written into my purse with no intention of filling it.

But it took less than twenty-four hours until I did.

---

One of the first big decisions I made was to get an assistant. I needed someone to help organize my life, send emails, keep me on some sort of track of normalcy as I waded through doctor's appointments, horrid insomnia, and another bout of declining health. This time I had no deadline, no manuscript to complete, no New York to return to, but I felt I had to act like it was all there or else I'd drown again. And with no real explanation of how this had happened all over again. This time I had to have answers—I couldn't lose myself again.

I made a Facebook post asking for an assistant I could modestly compensate in the LA area, and Zed was the first to respond. I knew him only vaguely as the younger brother of the one other Iranian girl I knew in my high school—she was a fellow Drama Club member and a year or two younger than me. I was gone by the time her brother attended our high school, but in his message to me he wrote that he was a fan of my work, so proud of all I had done, and he was in between jobs and looking to change his life.

*I want to get on a good track now,* he wrote. *I look up to you. And I know I can help you. It's what I do.*

It was unclear to me what that meant exactly, but at our first meeting I learned a few things about him: that at one point he wanted to be a writer but abandoned it, that he was a sort of physical trainer to local people but without any actual certification, and that he had some significant substance abuse issues but now it was all behind him.

I knew he was hired when he kept saying, "We got to get you off those pills."

"Exactly," I said. I was so relieved that someone else was also disturbed by it. "I was addicted to these things and now I'm on them again. A problem on top of another."

"Pills do all sorts of shit," he'd always mutter. "We'll get you off. We got to focus on that."

And because I had seen the dark side of pills it seemed fair enough to believe they were killing me. Getting off them was at least something to pursue.

Meanwhile I had Zed draft notes to various people I owed work to, help order the special mold-free cleaning of all my possessions (only one company in all Los Angeles did this for individual clients and it cost thousands), help me buy all sorts of wild things like the highest-grade air purifier one could purchase, and more than anything drive me around, mainly to doctor appointments.

"I've got to get better," I'd say over and over to Zed, betterness always feeling somewhat attainable around him.

---

The worst problem with the Klonopin was that it barely worked. The first night I took it was different from that night in Germany when my mother was visiting. It felt speedy, euphoria-inducing almost. I loved it as ever, but it was not doing its trick for insomnia at all. I was continuing to sleep terribly and yet continuing to take Klonopin because—and my psychiatrist agreed—to just come off it now after some weeks on it would cause too much havoc on my system. Better to just have a bit of it in me to keep things smooth.

And yet nothing felt smooth. Every day I began having those horrendous panic attacks that led me to take some Klonopin during the day, too, just so I wouldn't "withdraw," as I interpreted the panic attacks as signs of withdrawal. All the same logic and illogic of my summer of 2006 came flooding back. As my sleep grew worse, a deepening depression entered the picture. I became very irritable, finding everything from light to noise to smells to be intolerable.

One night I could not stop screaming after I smelled a chemical detergent my mother was using to clean her kitchen. It was as if the most intense gas was filling my lungs. I felt like I was drowning in its toxicity, and I blew up at my mother.

"See, you're trying to kill me! You brought me here to finish the job! I can't be around this stuff!" I wailed. I noticed my mother barely blinked an eye. She was already used to what I had become—someone beyond reason, someone lost in another dimension struggling to communicate with those she used to know.

It went like that for a long time. From April through August 2012, I had dozens of doctors, went to doctors' appointments nearly daily, and nothing came of it. I had theories, some supported by test results, some not; some fed to me by doctors, some by internet research; some by healers, some by MDs; some by my parents, some by friends; some from my gut, some by investigation: MS, ALS, PCOS, endometriosis, scleroderma, lupus, HPV, cervical cancer, ovarian cancer, hyperthyroid, hypothyroid, thyroid storm, dysautonomia, anemia, insulinemia, diabetes, Addison's, Parkinson's, Hashimoto's, candida, dementia, ketoacidosis, West Nile, and, yes, Lyme. We looked into everything.

None of the doctors spoke to each other, specialists were piled on top of specialists, each tugging and turning me in their own directions. They seemed as clueless as I was, my body a mystery they couldn't solve. I started to feel rejected by them, sensing their dread when they'd greet me, feeling the frustration in their bodies as they pored over yet another batch of bloodwork.

Meanwhile: losses. I lost out on things—an apartment in New York, one in Santa Fe. I'd been chosen to be a resident at the writers' colony Djerassi, and I told them a week before it began that I couldn't go. I was due to teach at a low residency MFA in Tampa, Florida, and just days before, I backed out of that, too—only after having a disoriented, insane conversation with the program director, asking over and over where the local hospitals were at the residency and how easy would it be to get medical attention there, trying to imagine how I would survive. People would write to me to get me to write things for them and I'd say yes, and then after trying I'd bow out. Students wanted me to work with them for manuscript consults and I'd say yes—followed by a quick no. It would take a second to remember myself, what I'd become.

My full-time job became my health, which was now a mystery illness that was hopelessly complicated by and tangled with addiction to psychiatric medications. I couldn't believe I was there again, after all the bad experiences I had had. Back to a life of pill bottles and pill cutters and days measured in dosages.

My emails at the time—to my mother, to friends, to doctors, to therapists—went like this:

> i think gabapentin is what is making me insane—rotting my teeth and making me so depressed and have all sorts of crazy symptoms and need to get on getting off of that
>
> think i've been drinking WAY too much water and that made my electrolytes loopy
>
> I just like to feel "covered" before bedtime! I want to be able to go out and know I can take something and be all good.
>
> I have sky high progesterone or did a week or two ago because I was prevented from ovulating thanks to a stupid doctor in Germany who thought it was the answer for PCOS
>
> Right now there are so many unknowns—I have a lot of "vein gurgles" (medical term) and weird skin stuff and not walking so properly.
>
> I took Klonopin for 3 nights last week, which is all it takes for trouble
>
> I have increased salivation and gum bleeding, weird heart rate and blood pressure and dizziness and disorientation.
>
> Sometimes I can't feel my hands
>
> if candida was detected in the blood, then is this sepsis? just

talked to a doc who said if it is detected in my blood then i'd be very very very sick (even more than now) and would have to be hospitalized

awful insomnia, abdominal pain and reflux and esophagitis and reflux these days but i've also had joint pain, spinal weakness, chills, fever, very strange headaches, memory problems and head fog, anxiety and panic attacks, rosacea, fatigue

I've realized my urine is entirely too alkaline

What is keeping me up at night or worrying me during the day is nothing but side effects and weirdness.

Rhabdomyolysis, they say, is on a warning for gabapentin. It might be worth looking into because my absorption of the drug this time around is what surprises me.

I'd really investigate cerebrovascular areas. It really seems right to me. I developed a stiff neck suddenly and severe nausea today (could not really hold food down) after a weird popping headache—usually pain does not break through the gabapentin.

Heightened confusion and now really can barely express words properly, with limp in one leg.

i have really bad dysphagia. could this be the ativan?! the neurontin? I am losing my ability to swallow, choking, can't get applesauce down even

i have a test that indicates i may have an autoimmune disease, sigh

These messages were rarely asking anything, often just stating something. I was inconsolable. It was hard to know what

I wanted out of all this. I wrote them as a person who could not be helped, who knew this, who could live with just being heard, a sign of still being alive somehow, perhaps.

I also wrote this, a letter to my mother, a week into our settling back into Glendale from the long journey in Leipzig:

> *dear mom, i am sorry for any trouble i am causing. i love you very much. i thought we could try one thing that i tried to ask the other night that embarrassed me–i wonder if you would consider sleeping in my room? i can keep your exact schedule. i feel really ashamed and unhappy at all the unhappiness i am causing. please know i do not want to be sick. it is my dream to be done with doctors. love p*

And for some time, she did.

---

Soon enough, though, relations with my family became fractured. A part of me began fantasizing about living at a psychiatric ward—it seemed logical, a place where I could get help, where people could tend to me, where no one would be too shocked at what was happening. But as long as I was under my parents' roof, that would never be an option. I began asking to sleep over at friends' homes, especially my writer friend Anna and her actor husband, Oliver. They were very Hollywood, newly married, and cheerful and carefree—my magical thinking told me that just by being around them, some of their goodness might rub off on me.

I met Zed at a Pasadena Mexican restaurant late one Sunday night, where I bought him several margaritas and finally got him to agree to drive me to the West Side where Oliver and Anna lived. They had agreed to let me stay with them for a change of scenery. He passed me over to them with some tense discussion at their Laurel Canyon home. "She really needs someone around at all times, she's constantly trying to get to a psych ward, hospital, you name it." Oliver and Anna chatted with me with worried faces, but also helped me get Ativan, from a friend of a friend, when I told them I couldn't sleep—we ended up in Venice at midnight scoring the bottle. They did this instead of committing me to UCLA Psychiatric, which I initially kept telling them they should do—but Anna had written extensively about her psychiatric hospitalizations and was totally against it, and Oliver agreed we should avoid it. I stayed with them for days, but at the end I felt too weird to push it any longer. Their happiness as a couple also started to make me feel empty—it reminded me I used to have that.

Eventually I made it to a psychiatric hospital, or I nearly did at least.

For a while I just kept going to my old hometown ER, Huntington Hospital in South Pasadena. This was my fourth visit to the ER room in a month. The main physician, Dr. Kalder, was always helming the ward those afternoons and he recognized me immediately, with a look of deep dread. After a few hours of nothing but waiting, shivering on a stretcher, he came up to me and snapped, "Why don't you do everyone a favor here and go where I'm recommending?" Where was that? I wondered. "Psychiatric."

I looked at Dr. Kalder and told him I did not want to be in a psych ward—when it came down to it, I felt in my body that it was not the right place ultimately—even though a part of me had always dreamed of its unknown underworld. I said at least I could do detox, but the social worker at his side refused.

The truth was I had lusted after the idea of detox for months at that point. I had brought it up with my old UCLA therapist from nearly the beginning, the first time I felt the Klonopin take hold of me again. I had become used to calling all sorts of rehabs in California, using their free consult option, keeping them on the phone as long as possible, explaining to them that I was under doctors' care and yet I was a drug addict. Because of the doctors *even*, I was a drug addict. Street drugs had never been a problem—it was always their prescription pills, always procured legally and with their consent.

Almost every rehab I talked to, they sighed at the word *benzodiazepine*, some also sighing at *Neurontin*. They never seemed that hopeful; they could never guarantee me any sort of relief. *No promises*, they always seemed to say, even at a place called Promises. They were also prohibitively expensive—on average, it seemed a rehab program would cost $30,000 for a month. My entire National Endowment for the Arts fellowship money, plus any spare change I had at that point after the many bills and lack of work, would go to a month of rehab, which might do nothing at all.

But my strongest hesitation came from knowing that it would only heal a layer of me at best—there was something underneath, whatever had gone wrong before the pills, whatever had made me sick in the first place. It was hard to know just which era of illness I was thinking of at this point—years of affliction weighed on me chaotically. All I knew was that it was time for someone and something to finally fix me and it didn't matter what.

I blinked at the harsh fluorescent lights of the hospital that by now I knew so well, like a California sunset.

The social worker was shaking his head at me, smug for a moment, then quickly putting on an expression of solid concern. "No detox needed. Psychiatric is more appropriate for you."

They left me to think about it and by midnight, I was ready to give up. I didn't know where else to go. I wanted to be well and if a psych unit was going to be the place for that, then so be it.

"Excellent decision," Dr. Kalder declared, without even meeting my eyes. I was a good student, but I had worn out my welcome. It was more than time for them to move me on.

My possessions were gathered and handed to me, and I was placed in a wheelchair. A security guard wheelchaired me through an underground tunnel that connected ER to psychiatric. In the unit, a team of nurses came up to me without a greeting and immediately went through all my bags. They took my cell phone and computer and keys and pills and shoelaces (anything I could use to harm myself).

I called my parents, who were past the point of mortification. "This is going to help me," I assured them, though by the looks of the place I was not sure. "We have to give something else a shot."

"You think you're crazy?" My mother cried into the phone.

"I think there is something very wrong with me," I said calmly. "I think there is something wrong in my body. But they think it's the mind. So let them do it their way. Let's see."

Silence.

I never made it to the actual psychiatric ward. I was in the holding area for three days, Memorial Day weekend came and went, waiting with those who needed to get medically cleared—mostly elderly demented patients—while physicians and psychiatrists evaluated me. They suggested antipsychotics if I didn't want more benzodiazepines—that same old Seroquel from that awful LA summer popped up again, like a bad bogeyman—and I refused. I decided to go through benzo and Neurontin withdrawal during that time, and it was not pretty. I did not sleep at all and finally caved in to Valium, the one benzo

I had not tried. Meanwhile they watched, took my blood pressure and temperature and heart rate over and over, not knowing what to do with me.

"I don't belong here," I'd whine to the nurses.

They'd assure me I did. "Nothing is wrong with you physically, you need to understand that," they'd say again and again. It was a line I had heard many times at this point, so it washed over me like water.

And yet in the end, I could not get "medically cleared" to go to the psychiatric ward.

When I didn't get admitted and my parents took me home, I knew it had gone too far. I promised them I would face this, that it was just anxiety. That meant ignoring heart palpitations, electric shocks through my limbs, piercing headaches, and unbearable fatigue, but like the doctors had all said, the mind is powerful. I had done this to myself, I began to tell myself.

"Good, let's just focus on good things now," my mother stroked my hair, trying to smile through tears. "You are okay. You are okay."

"I am okay," I said, matching her smile, matching her tears. "I am okay."

But I always kept a little part of me, like a single hair in a locket, this bit that never gave in and never believed it.

---

In high school, my senior quote was something I plucked out of Whitman's *Leaves of Grass*: *Through angers, losses, ambition, ignorance, ennui, what you are picks its way.* I was always someone who had a great sense of herself. *Most Likely to Have Her Own Advice Column*, I got in my senior yearbook. *Wild Horse*, I got from my mother as a teen. A little bit know-it-all, mischievous, outgoing, friendly, hot-tempered, smart, more grounded than I seemed.

If there was one thing I never had to wonder about, it was who I was. It was skipping that journey that I sometimes think got me to being a writer, storytelling being the ultimate confidence game: *listen to me, I have something to tell you.*

I tried to remember these things, as I went in and out of hospitals. One night at that Huntington's psychiatric holding den, a young nurse who had become fond of me was helping untangle a rubber band from my hair while saying, "What happened to you?" She wasn't talking about my hair, she said. And I knew that wasn't part of her script. I decided to tell her to look me up, though I immediately apologized for sounding like *Do you know who I am?* But I hoped she'd understand that in my own way who I was had value: novelist, essayist, journalist, academic. She came back after nearly a half hour with tears in her eyes and said one sentence before she gave me a quick hug and left: "My God, you had a whole life." I nodded and nodded, bobbleheaded, ecstatic, devastated. But it was the first time I had broken through a bit, been offered a mirror image that finally made sense no matter how simple: *a human who had a life. This is who I am*, I thought, *a real person, an achiever person, a person whose intelligence had driven her, not a train wreck, a child, a drama queen, a hypochondriac, a basket case*. No one is just those things, and sometimes no one is those things at all.

I tried my best to steer clear of ERs after that and instead I followed a writer friend's advice and began seeing her psychiatrist, Dr. Behzadi. My UCLA neuropsychiatrist was no longer working with me—he was too fed up, said I wouldn't follow directions, didn't know where to go next, and at the same time refused to give me a plan for getting off Klonopin. *You can just get off it*, he would say over and over, but I was convinced it was dangerous—and I wasn't ready to drop it. The last thing I wanted to do, in such choppy waters, was rock any more boats.

Dr. Behzadi was Iranian, too, flashy, handsome, littered his psychology talk with hip-hop lingo, saw many celebrities, and had an unshakable confidence in me getting well that I almost trusted.

"You are not going to be this person in just a few months," he said. "I can tell you that."

"How do you know?" I remember squirming on his leather love seat, my body feeling unlike my own as ever, wondering if you paid someone enough they'd tell you anything.

"I am a pro in this field," he said. "And the first thing I need for you to do is trust me. I know it takes a second. But I've seen many patients like you. Far worse even. And they get better. You will. Believe."

I believed him.

I especially believed him because he believed there was something wrong with me. "Look, I can't argue with what you are telling me. Some of this does not sound psychiatric. And we're going to get to the bottom of it. Because you're going to get better, remember?"

He immediately referred me to his partner, a medical doctor, Dr. Frank, whose website tagged him as a "VIP Medical Concierge." He also seemed to deal with mostly celebrities. Behzadi wanted me to get a clean bill of health from Dr. Frank, so he could prescribe me drugs and also to soothe my health anxieties.

Dr. Frank himself looked like a celebrity, or at least a doctor in a movie, short and trim, with flashy suits and trendy haircuts. He pored over my piles of paperwork and chuckled and snorted at much of it.

"We've gotta lot of work to do, don't we?" he said, winking at me.

"Do we?" I said. I was humorless those days.

"Don't worry. We're gonna run every test there is. We're gonna get you back to business. You write movies?"

"Books," I said. "Articles."

"No movies?"

I shook my head and pretended I didn't see that glimmer of disappointment.

Weeks into the Behzadi-Frank appointments I was still getting worse. And it was just outside Behzadi's office on Sunset Boulevard that I collapsed right between the sidewalk and the lobby of the giant skyscraper, falling to the ground after many sleepless nights, and got taken to Cedars-Sinai. It was only after a few weeks of seeing Behzadi, and he was very shaken, seeing me screaming on the floor when I came to. He called 911 and he also called Dr. Frank, whom I'd seen a few times at that point, and it was Dr. Frank who got me out of the ER and admitted into the Cedars' cardio ward.

I never found out why I was sent to the cardiology division, but I could only assume that of all the things wrong with me, maybe my heart was most suspect. My friend from high school, Nicole, was a doctor there, and she visited. She seemed unsure about me being there, telling me admittance was usually very expensive.

"But if they can't help you here, then I don't know where else. This is the best," she said, eyeing my vitals, promising to be back.

My mother eventually arrived from work, her face pale from panic, still trying to smile for me.

I tried to smile back. In the chaos of the hospital room, the beeping of monitors, the taking of my blood, the nurses yelling codes in shorthand, patients yelping and screeching, I heard myself say to my mother, "Maybe this is for the best."

I also kept saying something I had heard some other therapist or doctor say at some point, maybe in the psych ward: *Let's get to the bottom of this once and for all.* I was mesmerized by what "the bottom of this" could be, but I knew I wanted it.

My mother nodded glumly.

I got wheeled in my hospital bed to a fancy room on a high floor—much nicer than the ER, almost like a hotel, with stunning views of LA. I imagined this was how celebrities went to the hospital. The nurses all seemed cheerful and assured me that this was the best hospital in the world and I would get better.

"We are going to help you," one of the nurses said, as if to tuck me in.

I nodded, eyes full of tears, mostly in gratitude, I was realizing. I kept thanking her, thanking anyone who would be around to be thanked.

This had to be it.

But over the next days my life became even more of a nightmare. I couldn't sleep at all—not unlike the psychiatric holding area of Huntington, hospital admission meant you'd be interrupted at all hours for the measuring of your blood pressure and other vitals. The food was so unbearably bad that I couldn't observe any of my newfound good eating habits designed to address every potential health problem I could have. And I was torn away from my computer, my phone, any outside world.

The doctor presiding over my stay, who was apparently in touch with Dr. Frank, was a gruff, rude pulmonologist who was again fixated on my sleep—he wanted to try a different sleep drug each night. One night with Remeron, his top choice, which he piled on top of a dose of Ambien, I finally slept but woke up in a strange dissociative psychotic state where I could see myself in third person.

"I have lost it," I heard myself saying. "My God, it finally has happened, I have lost it."

For hours I think that's all I said.

He was not shaken by this or any of my behavior. He had me deal with a whole team of doctors: an elderly endocrinologist (who tended to visit me between midnight and 3 a.m., wanted to talk about my writing, wanted to give inspirational anecdotes, and believed nothing was wrong with me, but maybe a boyfriend would help), a neurologist (a very mean, stony, rude woman who thought I had MS at first but with the absence of brain lesions did not have any time for me), a cardiologist (also Iranian and went from being very worried about me to being altogether fed up with me), and a staff of young therapists-in-training who kept trying to convince me to take antipsychotics.

"We see this all the time," their leader said firmly. "Patients with cancer take them. It doesn't mean there is nothing wrong with you physically. It just lets us have a clearer path to that. So we can help you. So we can understand you. It doesn't mean you're psychotic."

I wanted to tell her, *But maybe I am.*

The days kept going on and on, and I could only measure them in the moans of the dying man next door.

If there was hell, this was it, my celebrity ward, the great glittery Cedars-Sinai, in the heart of glamorous Los Angeles, the jewel of the Golden State.

This was where I was rotting, this was where I was fading away.

In the end, only one doctor was somewhat sympathetic—a kind, gentle infectious disease specialist who was interested in my Lyme history. He watched my heart rate and read my file and told me if it did turn out to be Lyme, it would be what they call late-stage Lyme.

"Not what you want," he said.

"I don't want anything," I replied.

He said he would check the Lyme titers and I never saw him again.

By the end of my time there, the doctors didn't like me and wanted me out—this was clear. I'd been too demanding, not listening to them, sticking to my story as if it was fake, wanting answers, asking for sleep and real food. None of the test results were adding up to anything. The nurses began to avoid my buzzes, the moans from the old man next door died out when he actually died halfway into my stay, and my parents stopped visiting. The only diagnosis they came to was anxiety, plus a recommendation to get me off sleep medications and benzos, to be followed up with Dr. Behzadi and Dr. Frank. I found myself in exactly the same place I had started, except this visit cost over $80,000 and it was not certain what insurance would cover.

Back at my parents' condo, I was more alone than ever. I ditched Zed when I discovered he was a drug addict—he did a coke run with me in the car, driving into an East Los Angeles alley in spite of my screams in protest, getting a package from a wiry guy who counted his money more than once. Another time, he said he was too hungover to take me to a doctor and so I was rideless minutes before my appointment and had to take an $80 cab to the next city over. I realized I had been making excuses for him, thinking he was all I had, but I had to face that he was a mess and couldn't help me. Just another person who was better off thinking me dead.

"I can't have fucking drug addicts around me, Zed," I finally said to him on the phone. "Especially, you know, as I kind of am one?"

He was silent. I could feel the space between us grow hot with shame, his and mine.

"You fucked up," I went on. "You tried to take advantage of

me. But I don't blame you. You are an addict. Used to be and still are. But I can't help you. I brought you in to help me. Remember?"

Zed didn't say anything other than he had to go.

"So go," I said. "Don't come back either. Fix your life though. Fix yourself. And I'll fix mine." I nearly laughed at myself—how simple it could all sound when you were gone far enough from yourself.

---

Things were very bad with my parents. The more we tried to make them good, the worse they got. My mother began crying whenever I started getting into my illness, locking herself in her room not to hear me anymore. She was drinking more than usual, I noticed. My father also told me a few times that I should leave, that they were peaceful without me, that he wanted his happy life with his wife back. He seemed to back away from me every time I spoke and often left the room when I was at my worst. I'd lost them.

For Mother's Day, we decided to really make it a special day for my mother, since she'd been under so much strain because of my illness. For the first time in years, I had a large bouquet delivered to her office. Then that weekend we asked her where she wanted to go, and she suggested lunch at a Persian restaurant in Newport Beach and maybe a walk on the beach. We went, eager for a day where things could be normal, but while we were at the Persian restaurant I felt violently sick again. I had to keep going to the bathroom just to gather myself. People were looking at me, and during one of my trips to the restroom, an older Iranian man told my father they recognized me, *that she was a great Iranian writer, was she not?* He was so proud. When I came back to the table, I tried to finish my food in tears, tried to seem appreciative, tried to make it through the meal for my mother's sake, but her eyes were on me the entire time. Still, I rallied for the beach, for the ocean air, hoping things would improve.

As we were driving in circles looking for parking, a bad feeling came over me. Suddenly I burst into tears, and my family was alarmed and consoling me, while I was trying to push it all away for the sake of my mother and Mother's Day. There was a lot of tension in the car and I could tell my mother wanted to just go home. She was just going through all of this so she could say we gave her a nice Mother's Day, for our sake.

Then, as she was speeding around a curb, she suddenly hit something. A young man in a helmet leaped out and away from our view. A few people rushed over to him. We stopped the car and we didn't get too close, but just watched the biker, who seemed hurt but it was unclear now. My father tried to go say something, but my mother stayed in the car and snapped that the man fell on his own. My father came back and for some reason I tried to tell onlookers *I am very sick and need medical and psychiatric help*—as if my misfortunes could excuse us—as I backed into my mother's car. Like a bad dream I don't know how or why we went off on our way.

If we even did, that is. Around this time, I also found myself imagining many things. My own perception could not be trusted either, I knew even then.

I knew I had to do something more to help myself, or so I could get myself to the point of helping others help me.

I knew I needed community, if I couldn't get that support from family.

Even after we had driven off in the car, my mother denied the accident ever happened and quipped cheerfully that it was Mother's Day and let's not ruin the day. For weeks and months after, I worried about what had or hadn't happened. My mother, my big ally, became another person not to trust, and in some ways my father too. Like Zed, everyone was bound to become untrustworthy.

Where to find community, my people?

Not too long after that incident, I got involved with Point of Return, an organization that helped people get off psychiatric medications, at the urging of Firoozeh, who had googled them and now believed my only problem was benzo addiction. I called them and I got the founder, who immediately looked me up and treated me as a VIP. She kept assuring me, "You know, we handle quite a lot of celebrities here so you'll fit right in with us." They sent me their supplements—whey powder and sour cherry juice and all sorts of other natural products that were said to have superpowers, at very high prices—and gave me a daily schedule and told me I needed to call them all the time and stay in close touch for support and medical monitoring. I was suddenly talking to a network of doctors all over the country, who were giving me all sorts of advice without seeing me. Their main advice: get off the meds. I wanted this so at first I didn't question them.

They also sent me their book, their founder's self-published memoir. I read through it, inspired, but I started to find some things familiar. It reminded me of snippets I'd heard about Scientology and all the emphasis on purity and detox and drug-free lifestyles started to click for me. I realized that they might all be Scientologists—that all through it there were codes and analogies that pointed to Scientology and that certainly an anti-meds group might be linked to Scientology. I got the courage one day to ask the founder on the phone point-blank.

"I mean no disrespect, but I have to ask, as I've researched it some—are you all perhaps affiliated with Scientology?"

I heard something between a sigh and a groan. "It's really

inappropriate to ask about this. I want you to think about why you are asking, why you'd bother me with this." And she hung up.

I realized I could not trust her, either.

(I never heard from her directly again but they still keep me on their mailing list, no matter how many times I've tried to unsubscribe.)

I started calling doctors on the phone—hospitals and detox centers and psychiatric wards, looking for answers again, deeply wanting connection with someone who could help me. Every call seemed to be another dead end.

Then came the worst of those exhausting days. It was a day I knew I could not get through, when the possibilities seemed hopeless, all but one. I was burning hot and sleepless, my thoughts bobbing over to suicide again, but I knew I owed it to people I loved to try absolutely everything before that. I watched the clock as if it was a time bomb and decided to do the one thing I told myself I wouldn't bother with, to address the one marginally nagging concern that still crept around in the corner of my mind at times: Lyme. I decided to call Cedars again.

I never quite forgot the infectious disease doctor and how he'd said he'd test the Lyme titers, but I'd never heard anything more. I recalled that he was the one doctor who actually listened to me during my stay at the hospital, so it was worth a shot, though I imagined I was negative for Lyme as they never reported back on that test. And even if it had been positive, I wondered what could actually be done about it, since I had tested positive for Lyme in the past and that hadn't amounted to anything. Still, I dialed with a great deal of fear, which was tempered in part by the knowledge that it would be highly unlikely that I'd reach him—that those doctors came in on rotation and why would there be an infectious disease doctor on hand to take my call, let alone the same one who tested me.

I would have taken it as a sign if I had still had some grasp of hope, but I actually reached him. And not only that but when I started to remind him of who I was he interrupted me: "Of course I remember you." I was shocked. "But how could you not be better yet?" he asked. Then he looked at my files and said that it turned out they had never tested the Lyme titers—the pulmonologist had ruled it all psychiatric and they'd had to turn over the room. He was very sorry. "I can't say for sure what's wrong, but looking into the Lyme route might be the right thing. I can recommend you to my good friend, a doctor in Malibu, Dr. Mills, who does Lyme cases." I wrote down the number because this had been the closest I'd gotten to a miracle in ages.

Around the same time, I was also being encouraged to look into Lyme by a young man in Santa Fe, whom I'd met through Firoozeh and who had been urged by her to stay in touch with me. He was convinced I had Lyme like his brother, and we began talking once a week by phone. He began to suggest I come to Santa Fe again and stay with him, saying he was sure it was Lyme, and he was sure I could get better there. I didn't know what to say. From all I had heard, the Lyme road was long and frustrating, so a part of me hoped for something simpler. But then it couldn't be much worse than my current state of unknowing, so between his encouragement and the recommendation of the infectious disease doctor, I decided to investigate further.

I reluctantly called the Malibu doctor and he immediately seemed convinced I had Lyme.

"I have little doubt, to be quite frank, but we shall see," he said.

I reconnected with Zed out of desperation, because both my parents were at work and at this point were utterly fed up from driving me, and so he drove me to the doctor's office one morning.

"Here we go again," he muttered, seeming jittery. I wondered if he was still on drugs.

"I want to say I have a good feeling about this but I don't really have feelings anymore," I said.

Neither of us laughed. There was never a quieter car ride.

We got there early, Zed speeding, me thinking about telling him to slow down but deciding against it. The office was nestled in what looked like an alcove of surf shops, and inside was a mix of bourgeois hippy surfer chic and Malibu housewife glamour. The doctor was much older than I thought, but he was the epitome of California mellow. He had a secretary who looked and acted like his granddaughter in a neon tube top and matching flip-flops, who used words like *stoked* and *psyched*. He drew my blood himself and did so quite badly, missing my veins a couple times. He sent me home with megastrength probiotic samples. He seemed pretty sure I had Lyme.

"How do you know?" I asked twice.

"I do this all the time," he said. "It's all over this place. The rest of LA hasn't seen it like we have but over here, it's everywhere. I know what Lyme looks like. I can see it in your face."

I looked at my reflection in a giant framed photograph of waves at sunset. I could only make out my hollowed cheekbones, big unblinking eyes, the huge shock of frizz that was my hair now. Malibu was a mostly new area for me, a place I'd only gone to for short beach excursions, the last place I imagined would be where I'd seek a medical practice. Still, something about the experience felt right in a way nothing had in a long time.

In the car, Zed asked me what I thought.

I shrugged. "I'm not sure I trust anyone anymore. It's been a while since I've trusted anyone."

---

I was to go back to the doctor after a few weeks. He was sending my labs out to IGeneX, the premier Lyme lab in the US. It would take a while for results, and over $2,500 for the two types of tests he wanted.

Now it was just a waiting game. And as I waited I just tried to survive. One of the worst symptoms I had was dysphagia—a word I hadn't known before but suddenly knew well: the inability to swallow. At first it felt like there could just be congestion in my throat, perhaps an infection of some sort, but soon I realized it was chronic, and at a certain point I realized food was not going down. As if there was some kind of blockage, I'd find food I'd chew would want to fall back on my tongue. I'd concentrate so hard on swallowing there would be tears in my eyes from the strain. I started to realize that I was losing my grasp of consciously creating that movement. Could it be that I had forgotten how to swallow?

I sat on the phone with friends and had them slowly read me internet descriptions of swallowing—quite a miracle in itself, as it required the perfect orchestration of so many nerves and muscles—and I would try to focus on getting all my own nerves and muscles to cooperate. I started to seek out liquid foods, but even those were a challenge to get down. I tried throwing spoonfuls of applesauce down my throat with my head tilted back and just hanging there, praying. I'd sit up and blink and think all was well, until I'd feel it again.

I reminded myself I was still waiting on some tests. My mind went back to old solutions for survival, like working, but I couldn't imagine shopgirling at this point. On a whim, I offered my services to a Middle Eastern poet named Leyli, whom I only vaguely knew—anything for a distraction. She had been posting on Facebook and had recently invited me to some parties at her house that I had been too ill to attend. I decided to message her. I had realized I needed to get out of the house fast and do some kind of work, even if it wasn't paid.

*I can do anything, anything you want*, I wrote her. *I mean, I will try to do it. I will try to do anything. I'm very sick. But I think seeing you will be good for me. I have historically been smart and hardworking.*

She told me she felt heartbroken by the urgency of my message, its desperation and its emptiness. She had a great deal of career anxiety and had two books coming out herself, but she said she saw me as successful, so she felt it could be helpful to pick my brain. And on my side, I felt it would be nice to exercise my brain to think about literary matters again.

*I see no reason why this can't be great for both of us*, was her response.

Meanwhile, she had just gotten through a divorce and lived in a stunning cottage in the heart of Los Angeles, a sort of Secret Garden–style sanctuary with a separate back house that she kept offering me to live in. I went back and forth about it; her fairy-goddess platitudes grated on me at times, but it felt good not to be at home. She also had begun dating an elderly Palm Springs hypnotherapist, her first shot at online dating, and she wanted me to help him set up a law office in LA, too, in exchange for free hypnotherapy.

"This could be exactly what you need!" she exclaimed, always twirling a bit in her long skirts. "William is the best and he might be really able to get at the problem, whatever it is! He's very expensive so I'd jump at this!"

It was easy to say yes to anything when I was beyond the point of thinking anything mattered. So I sat with William in the back house one morning, the sound of birds distracting me, as we stared at each other. His old healthy tan face, my young broken pale face. I did more than one session, though nothing came of it, and it turns out I wasn't much help to him, either.

Meanwhile I had to watch these two people, old enough to be my parents, act like crazy kids in love, while I sat, a shivering trembling aged husk of a girl. They must have noticed it, too, as they decided I needed a boyfriend and recommended William's brother, Arnold. I assumed he was, like William, at least twice my age. "Age is a nonissue in this material world!" Leyli would insist. "I don't seem old to you, do I?" Still I tried to dissuade them, and we didn't talk about it further, until one evening Leyli invited me to her house for dinner. When I got there it was clear it was going to be a party—with Leyli and William, and his brother, Arnold, and me. I was being set up.

"Relax, you look lovely!" Leyli practically sang.

I looked down at myself. Same thin cotton shirt, same long skirt made of sweatshirt material, the only clothes that didn't feel abrasive. I'd started to have neuropathy, which meant at certain unpredictable times of day I'd feel like my skin was on fire.

"This is crazy," was all I could say.

"He's a little old," she told me as she set the table, "but he's so young at heart and plus you know how I feel about age! Numbers are nothing!"

William beamed at her in agreement.

I nodded slowly. It was just a night. Everything was a bad day and so I could do this one bad night.

Arnold arrived, and he was in his sixties, as I suspected. A sort of construction tycoon (though it turned out, also slumlord), he was one of those leathery-lizard types you might find in Phoenix. He was off in every way, awkward and loud and brash—and he had never really had a steady girlfriend. While

William knew how to smooth-talk and had the air of an old playboy, Arnold was rough around the edges, always erring on the side of rudeness.

He immediately liked me. "Not bad, not bad," I heard him mutter with a big smile at his brother at one point.

I wanted to scream, *I'm right here!* but I wasn't sure he'd care.

I excused myself at one point, clutching a pill bottle, and snorted an Ativan in the bathroom to make it through the rest of the dinner. It helped. I vaguely recall laughing at a joke or two of his.

If I had been myself, that would have been that. In my mind, Arnold was perched on one of those carnival game booths, the one where you aim a ball at a target and if you hit it, the man falls into the bucket of water. That's where the real me would have had him, but I was far from myself.

Instead, for the next few weeks, we sort of dated—if you asked him, that is. For me, it was getting him to drive me to doctor's appointments and then in exchange for his service, having dinner with him or even a few times making out with him.

"I like you and I'm going to be there for you," he'd say. "You need me and lucky you, I'm right here."

I rarely responded, I rarely said a word to him.

Around this time, a friend from Baltimore grad school, Andy, was house-sitting in LA as he got over a bad divorce, and we reconnected. Like with Leyli, I tried to offer my services, but instead our time together became centered around writing dates at various libraries, with long breaks over food. Back in grad school, we both loved to eat, but now it would nearly knock me out to try to keep up with him—I had so much anxiety about just getting food down my throat. I also barely knew how to work anymore, which meant mostly just staring at my laptop for hours at a time. Andy was one of the few people I felt safe around, an upbeat character from my past life who had only known me in the best light—until now. His fresh worry at seeing me this way reminded me of my life before and seeped into all our interactions.

Meanwhile, Leyli set me up with her friend, a Sufi healer in San Francisco. Soraya was in her sixties and was supposed to be working on a memoir, and she wanted to heal someone in exchange for ghostwriting. It seemed a perfect opportunity. I spoke to Soraya and she had a very out-of-this-world psychic mystical vibe that normally I'd never buy. And yet, in the grips of this illness, it hooked me.

"You are not only going to get well, you are going to soar! Remember birds? What do they do?"

I was silent, expected her to answer her own question.

"Well?"

"I don't have an answer," I said. "I guess they fly."

She laughed. "No, my dear, they are building heaven! For us!"

Somehow she convinced me to come see her, that this was meant to be and we could help each other. Somehow I agreed.

I went to San Francisco with Andy. He said he was going to visit friends, too, but he maybe said that just so he could help me without me protesting. I was nervous just showing up at this weird woman's house in the Marina District, but Andy told me it would be fine and that he'd be nearby if anything went wrong.

Soraya seemed more broken and tragic in person, more needy, a bit of a wayward lost wanderer looking for a way out. It turned out she had been very wealthy once—married to a very rich man—and had lost it all in her own health crisis. That first evening she placed in front of me a few tiny bowls of food: a grain I didn't recognize, an orange powder of some kind, and some dried fruit. She brought me a fragrant tea she called *the empress of healing*. She watched me pick at the bowls and sip the tea cautiously. "Color is already coming to your cheeks, you know," she said with a big smile.

After some time, she removed the bowls and presented me with a giant stack of papers, bound with three rings.

"Here is my baby!" she said. "I mean, our baby!"

I read her memoir over the next few hours. I skimmed much of the beginning, but eventually I could not skim. I read and even reread parts to be sure. It seemed to me that much of the memoir was untrue. The stories and dates didn't line up. There were claims that could not be verified about a friend of hers from childhood who had died and this person's role in the Holocaust. There were all sorts of other minor details that seemed fabricated or fake, and I felt the work was unpublishable.

When I approached her gently about my concerns, she got very defensive and we ended up having a yelling match.

"I'm a writer. I can't put my name on something that I feel is untrue," I tried to explain.

"No name needed! It's my name! And you write fiction—you make things up all the time!"

I groaned. "Is this fiction to you?"

She shook her head.

"Well, then, it's not right. Memoirs are supposed to be true."

She was indignant. "How dare you say that!" But she didn't deny making up the details I pointed out.

"I should probably just go," I said.

"No, we can fix it!" she insisted.

But I didn't know how to repair her fabrications, and no promise of healing could convince me to stay.

Before I left the city I met up with Jacob, who lived in the Bay Area now. He had taken another year of leave after the stress of our split and moved there to be near his brother. He was having his own crisis. We took a drive, went on a walk through town, and had dinner in a Whole Foods café. He said he wanted to leave philosophy and become a magician—back when we dated he loved card tricks with a passion—but he was also thinking of just working for Starbucks. He seemed so lost. He looked at me as someone lost, too, feeling terrible as he saw me struggling to swallow a meal.

We managed to have some laughs over our sorry states and exchanged hugs. The next day he called me, a bit shaken up. It was hard not to feel that we needed each other badly. And yet we had lost each other. We made plans to see each other again in the coming months, and when that didn't happen, we made plans to see each other again months later . . . and again and again for years. Eventually we lost touch.

That was the last time I saw Jacob.

---

When I got back to LA, I confronted the fact that my work life seemed to be out of the picture—I was not ready for the madness of the world, and it was certainly not ready for me. I committed myself to slowing down again and I realized that meant more waiting. The Lyme test results still weren't in from IGeneX yet, so to pass the time I let my mother coerce me into going to the gym with her. I was horribly skinny, but she had always been a gym member and she thought it would help my moods. We walked into her grim LA Fitness branch in Glendale—which was full of only meatheaded men—and they tried to convince me to be a member.

But Josh, this one personal trainer, managed to connect with me. He told me his story, that he "too" was a pillhead and turned his life around, after losing all his money and his job. The details all seemed generic to me, an after-school special, the usual stuff, but somehow it soothed me at that point to hear anything mirror my own demons.

"The gym will help, but more than the gym, I will help. I'm gonna be there for you all the way, don't you forget it, sister." He had that golden smile, that all-American confidence that could sell you anything. I recognized this. So many men had tried to fix me; so many men were convinced they could help. What was one more.

And like so many of the others, he recommended another man who could supposedly heal me. During one of our sessions, he mentioned a Dr. Wayne in Malibu who had saved him—a healer whom he'd be happy to pass on to me. "I'm not that guy, not a crystals guy, but this guy did it for me," Josh kept insisting.

After training with Josh a couple days a week for some weeks, with few results, I asked him for Dr. Wayne's information.

"You're welcome!" He slapped me hard on the back as I walked out of the session, Dr. Wayne's information scribbled on a corner of an *US Weekly* page, the numbers across a starlet's tan bare shoulder.

I called. They could see me immediately.

"How does it work? How much and what do you do?" I asked.

"We prefer that you come here and we can discuss it all," a woman with a flat affectless voice uttered, as if it were scripted.

And what did I do but agree. I had all the time in the world at rock bottom.

It turned out to be another beachfront office not far from my Malibu Lyme doctor, Dr. Mills. And it also turned out they were surfing buddies. But this place was not a relaxed surfer den. Dr. Wayne's products were advertised all over the office, with big posters that boasted makeshift scientific charts. He had an overly made-up Russian receptionist who looked like a failed model, and her only mode seemed to be dismissive. He himself looked like a cult leader with his crisp black turtleneck and bald head, and he had a staff of cult followers who had titles like *cranial therapist* and *naturopathic healer* but who offered very liberal takes on those things.

Dr. Wayne ran some tests and applied some remedies: electrodermal biofeedback machines, muscle exams, blood chemistry evaluations, computer laser analysis, advanced light therapy diagnostics, electrical stimulation, cold laser, traction bodywork, and advanced chiropractics. He did it all very gruffly and very confidently, and in the end said I may have Lyme but my real problem was brain damage from my two car accidents. And drugs.

"Especially heroin," he said, staring me right in the eye.

I told him I'd never done heroin.

"You might think you never have," he said, unfazed. "But you have."

I felt frustrated. "How is that possible?"

"Have you done drugs?"

I nodded slowly. I could see where he was going.

"Ecstasy?" he asked.

I nodded. Immediately I was bombarded with flashbacks to the nineties at Sarah Lawrence and the warehouse parties and raves and clubs, and the jokes about whether you'd rather have your E cut with speed or dope.

He nodded back, a bit pleased with himself. "Well, it's often cut with heroin, don't you know that?" That stuff was lodged in my spine, he told me. "It's all over your tests." He said he and his staff could help me, but I needed to be dedicated. "Can you be?"

I nodded eagerly. "I'm willing to try anything to get better," I said, my voice cracking through the words that I meant with all my heart.

He led me to his remedy store. There he passed me bottles, jars, tinctures, tubes, balms, sprays, oils, crystals, supplements, you name it. He sold me hundreds of dollars' worth. And he even gave me the number of a young Iranian girl whose life he saved—and she actually called me gushing, sounding like an infomercial: *The man saved my life, I tell you!* I bought it only because I had no other choice.

He became another one of my many doctors.

Each time I went there, I saw him as well as his staff, a circuit of odd folks. One woman gave me cranial therapy with a laser, which was supposed to zap the brain damage out. Another did a violent style of chiropractic work (he was the fourth chiropractor I went to that summer) that made me leap out of the chair, my spine on fire, but apparently it was nuking the residual drugs of my past. And then I had these sessions with Dr. Wayne where he would tell me everything from *Santa Fe is full of sick people* to *kale will make you sick* to *In-N-Out is the best food*. Over and over he dissuaded me from the Lyme investigation path. *Don't go there; people spend their entire lives stuck in that one*, and it's the one thing he said that felt completely true to me. *You need to stay here in LA to get well.*

I'd nod, I'd go home, I'd come back.

And it was sometime around then that I realized that actually, the key to me getting better was to get out of LA.

---

It was not going to be an easy feat leaving Los Angeles, though for months and months I dreamed of it. And I knew New Mexico to be the next logical step, with the only step after that being the only step left in life for me: the return to the home of not my childhood but my adulthood, New York. But for the time being, the purple sunset and high desert air, the chiles and the dreamcatchers of that in-between home, New Mexico, lingered as the home of my healing.

Still, it was a challenge to extricate myself from the sticky trap of my parents and wayward friends and endless diagnoses, the place of a dull constant illness and addiction and death wishes. For one thing, I didn't know if I could survive without these people, this place—I could barely swallow food, I could not drive, I could hardly get through a night's sleep. I often gripped my mother's or father's arm, begging them for a ride to the doctor. How could I be on my own again?

Plus, every once in a while, the city would earn its keep for me, for just a moment, and I'd imagine it was good for me. That summer brought just such a moment. My friend, an Australian writer named Nathan, wrote me—he was in town from Sydney, though he tended to live most of his life at writer residencies all over the world, so he'd pop up nearly everywhere at some point or another. He invited me to a Fourth of July party in Studio City.

Fourth of July was never a favorite holiday—I'd always hated it and its culture of bland barbecues and boring fireworks and blind patriotism, but I got my parents to drop me off, both of them thrilled I wanted to go to the party, any party. They wanted to see me alive again, whatever that meant, and this seemed to them an indicator, even though I was quite unconvinced. I walked into that patio filled with dread, facing a small group of people I didn't know—Nathan's Australian writer girlfriend, Eileen, a writer named Dorrie, another writer I can't recall, and someone I actually did know: a filmmaker named Michelle, whom I met at Yaddo that same summer I met Nathan. Nathan and I had pretty much only exchanged emails since then, and I felt that strange closeness to him and Michelle that somehow comes from spending a month together at a residency. They felt like distant relatives—or, at the very least, supporting cast from a better time.

"What can I get you?" Nathan gestured toward the well-stocked patio bar, and I recalled the last time we were together was one of those endless drunken nights at Yaddo. Nathan and I were in fact almost never sober together. We could both hold our liquor, that was for sure—once upon a time, at least.

I shook my head and he seemed to get the message. I had written him an email about my condition, but I doubted it had been clear to him until now, when I, a healthy drinker, former professional drinker in fact, was not going near alcohol. As I was the only nondrinker at the party, the subject took a downturn to my health.

"How horrible, a mysterious illness . . ."

"Well, you look great, in any case, not ill, not ill in that way . . ."

"How terrifying—I'm sure you'll get to the bottom of it soon . . ."

"I actually wish I had the good sense not to drink, too, I've had all sorts of conditions myself . . ."

Michelle in particular seemed sympathetic to my condition. "I haven't been myself since I moved to LA. I constantly think something is wrong with me. I mean, not that *you* think something is wrong—something is clearly wrong, or not clearly. Or you know what I mean. All I mean is, I can only imagine, if I'm healthy and yet I feel so bad too."

Dorrie also immediately sat next to me and told me she had been through some hell and was just barely climbing out of it. "Look, I don't often say this to people, but if you ever need a friend in any way, I am here. I mean, I'd be happy to do whatever you want. Go to a doctor, take a walk, have tea. Seriously. I think in some way I get it."

The party went by with little effort and I went home full of hope that maybe in that network of people there was some miracle way out. I was heartened by Michelle and I also thought of Dorrie and her offers. I was consoled to know Nathan was in town for a while longer too. He was not much of a driver—in fact he confessed he'd never driven himself in the US—but he promised to drive me around the next day or anywhere I needed to go for days to come.

It occurred to me that I should take him up on it, especially when a Beverly Hills gastroenterologist had called me for an urgent second appointment, after some test results had come in. My gut told me Nathan would not only be a good distraction but an essential support. I asked him and he said he'd meet me there and then we could go out and eat. Something in his voice sounded off to me though.

"What's wrong? If you can't do it, I understand. I don't want to bother you."

"No, it's no bother," he quickly said. "I want to. I really do. It would be good for me."

"Good for you?"

"I need to get out and get away for a moment. I've had a problem."

He was mysterious about it but I figured we'd get into it in person.

On Santa Monica Boulevard he flagged me down in his rental and we parked around the corner from the old shop on Rodeo where I had played shopgirl, when I first found out, sweeping at the crack of dawn, that an editor in New York was interested in my writing. Nathan seemed barely there, as I recounted that story.

"You're not anxious about my appointment, are you?" I asked. At that point, I certainly wasn't. I was no longer holding my breath, was no longer thinking that whatever was wrong with me would be discovered.

"No, no," he said. "We can talk about it later."

And so we entered the gilded lobby of the brash gastroenterologist, a young man who seemed determined to be the one to figure this all out. He had sent me for all sorts of tests.

And he had results. He seemed so proud of himself when he announced with a tight smile that one test had come out positive, that something indeed was wrong.

I barely blinked as Nathan squeezed my hand.

"Your autoimmune—it came out positive for scleroderma," he said. "Highly positive, I might add. We have it now."

I looked at Nathan and looked back at him. "What even is that?"

It was something I had not come across in my research. An autoimmune disease that seemed to attack the skin. It was bad. Later at home, I'd google and then google-image it and discover that such a diagnosis meant I could be turning into one of those lobster women of health-site nightmare photos.

We walked out of the office numbly. I wasn't going to see that doctor anymore—for him it was an incidental finding, as he was a gastroenterologist. He sent me with the name of a rheumatologist and his best wishes.

"I hope you can now rest easier that it's solved," he said.

I was not so sure.

Outside in the harsh last hour of full sunlight, that summer sun of Los Angeles that seemed to beam so recklessly, so heedlessly at every sorrow of its denizens, I burst into tears long and hard into Nathan's chest. He was for once wordless, and eventually I could feel his body heaving gently with mine, sobbing along with me, but with something more than sympathy pains.

We sat on a bench in a small park behind the doctor's office, adjacent to all the blaring glitz of Beverly Hills, in the shadow of the golden evening. I had nothing more to say, so he said he'd share with me what was going on with him.

He told me a secret about something in his life that had gone horribly wrong or so he thought.

"I feel like I've destroyed my life," he said. "I haven't been able to write in years."

I took a breath. "I relate to that completely, even though my circumstances are all different. But I feel like I've destroyed my life. I haven't been able to write in years, either."

On the long ride home, I thought of the two of us, laughing over bottles of rosé late into the night at the Yaddo mansion.

As if he could read my mind he said, "Things used to be so different for us. Look at us now." He was nearly chuckling, his eyes still wet and flashing.

"At least you get something past now," I said. "What about me? You left me behind."

"What?"

"You left me behind," I repeated, unsure of what I meant, on a numb sort of autopilot.

"Who left you behind?" he asked.

"Everything," I said, in what would be the last time I saw him for many years. "Everyone."

---

Meeting Dorrie and taking her up on her offers turned out to be the best decision I could have made. After Nathan was gone, deep into that endless summer of 2012, I saw Dorrie at least a couple times a week. She not only became my source of rides, but also was the person who would sit with me at meals, who I'd have writing dates with (we were both working on our second novels, or trying to at least), and occasionally she'd even get me to go to literary parties and readings.

"We're going to be a team, you and I," she'd tell me, and I'd believe her.

Slowly, Dorrie got me out of feeling like an alien. And she was there for many hard times too—emergency trips to doctors, including one morning in particular where my mind was so out of it that I couldn't tell if I was deep in benzo withdrawal or I'd overdosed. She even took me to an AA meeting, which she went to all the time since being sober, and the conversation with the small group of women held some light for me.

"You are going to come out the other end," she said. "I'm going to make sure of it." I had heard that before, many times. But this new friend of mine, so frail, so afraid of life herself, seemed so sure of it that I trusted her.

Dorrie made the bad experiences livable by laughing them off with me—she'd pull me out of the well of their pathos. We'd laugh about Arnold the Slumlord, and Soraya the Sufi Healer Quack, and Zed the Drug Addict Intern, and my impossible parents. We'd even laugh about how I was supposed to be a Lobster Lady of scleroderma, if I wasn't a Tick Lady first.

And so it was at lunch with her one early August day when Dr. Mill's Lyme test results came back: positive.

"Hold on, Doctor." I stepped out of the café. "Can you please say that again? I want to make sure I heard you."

He said it again. "It was positive. As we thought."

For a moment, I could not say a word. It felt like something that had been ticking inside for ages had stopped. I felt the absence of a certain tension, such a release that it felt nearly violent.

"Are you there?" my doctor asked.

"I am here," I said. "Somehow."

This time it did not feel like the scleroderma diagnosis, an impossible fit that had been pushed on me. This one made sense, deep in my gut. It was still monumental to hear it: an ending to my story, a beginning to another one.

Dorrie immediately saw it on my face. "Lyme, isn't it?"

I nodded vigorously.

She hugged me the way you'd congratulate a friend who'd won some award. We smiled at each other through tears.

On the ride back home, I felt like a stranger in my body, but this time I was a different kind of stranger. A traveler, or tourist. Someone who would one day look at this and maybe not even remember it. That was the closest I came to optimism in all those months. Could this really be the end of my road? Or was it a beginning? Was there any guarantee it could all end well? After all this now we had a diagnosis—one clearly more definitive than the one in 2009 that Dr. E had more or less dismissed—but what could recovery mean? I couldn't even imagine it. All I knew was that suddenly after those few words on the phone I felt reborn, like my life had begun again. I had no idea what it was going to look like from this point on, but I knew I at least had a chance at one—and it had been quite some time since I had thought like that. Just maybe, I was going to live.

There was much to be done. Dr. Mills immediately had me on a rare Canadian antibiotic plus doxycycline, and now the biggest problems for me were getting to him all the way up the Pacific Coast Highway in Malibu, a good hour from Glendale, and affording him, which even he admitted would reach deep into five figures at least.

Dorrie brought up fundraising one day when I had taken ages trying to put together the most modest lunch order given my anxiety about money. "You need to eat well—you need to get well and that costs money. You've got to ask for it—people do these days."

I was in my final days of being covered under Jacob's insurance, the one condition he'd agreed to in our split—not that Lyme was really covered anyway, as these expensive tests had immediately proven to me (and Dr. Mills had warned me that little else would be covered along the way). I had no other choice.

"There's only one thing left to do." She helped me log on to a GoFundMe site and we put together a little paragraph, and I uploaded one of the many sick-girl selfies my phone was full of.

---

In less than twenty-four hours we had over $2,000, and in that first week we were over $9,000. It helped that everyone spread the word on social media, especially my former students and members of the literary community who had also tipped major literary blogs and the *LA Times* book section on to it. I started to receive hundreds of emails a day, many of them including donations. Money, anecdotes, condolences, well wishes, love. Everyone had known something was wrong, but no one had known what.

Not only did I badly need the money—the fund-raiser reached $18,500 total—but I needed the support more than I knew. The highest donation, at $500, came from a woman named Sydney, whom I had befriended on Enders Island, where we'd hold our residencies, in my final semester at Fairfield University's low res MFA. Sydney was a San Diego transplant, who looked like what you'd imagine Barbie looking like at fifty. She was always in a thick mask of makeup, hair frosted like an eighties video ingenue, her thin tanned body (she taught aerobics on top of freelance editing) draped in expensive jewelry, wearing something a bit too tight or short or sparkly. I never imagined we'd get along but she was hilarious, very wry and irreverent, and now over a half a year since we met, she was my prime donor.

But that wasn't all—Sydney emailed and told me to call her, saying that she'd like to help me further. She asked me what could help me the most, and if I could have anything, what would it be at that point? I realized I had not thought about this at all. My own desires felt lifetimes away. I finally answered, "I'd love to get out of my parents' house and maybe to Santa Fe." I could stay with this guy for free, I explained—the guy whose brother had Lyme, one of the few people who urged me in that direction. I'd have support, my stuff, and old friends.

Sydney told me to come to San Diego and said she'd drive me to Santa Fe, one of her favorite places in the world. It turned out she had gone to college in Albuquerque and still went to Santa Fe often. Since she made her own schedule, it would be easy. So we arranged for me to come to her at the end of the month, and then take a road trip there, for me to at least try it out. I could always return to Los Angeles.

Daniel, the twenty-eight-year-old woodworker Firoozeh had connected me with, the one with the spare room and the brother with Lyme, was elated. He had no problems with housing an almost-complete stranger. *I redid the room especially for you,* he'd caption a text with a photo of an amazing rustic southwestern room.

I made an appointment with the one and only Lyme doctor in town, Dr. Canfield, an integrative medicine doctor known for merging Western and Eastern traditions and taking on the toughest cases. Who cost $280 an hour.

I deposited the donations.

I told my parents I was leaving.

I packed.

I slowly and surely disconnected myself from LA, my old home, the city I was raised in, but the place I thrived least in. I swore I'd never live there again.

In the meantime, Michelle, the filmmaker from Yaddo and Nathan's Fourth of July party, and I started hanging out. She had started a new documentary timed to come out with the new *DSM*, about the rise of mental illness and the overprescription of psychiatric medications. I realized she was fascinated by the mental health and meds part of my story—that throughout this illness I identified as a drug addict, too, most of my days revolving around gabapentin and whatever benzodiazepine I was on at the moment. She wondered if she could come along on the road trip and film me. I could not think of a reason why not, and I thought since I only knew Michelle and Sydney somewhat, together they could create the feeling of knowing someone more completely.

The hardest person to leave, harder than my parents, harder than any old lingering Los Angeles friends from my past, was Dorrie. But she maintained it was her own plan for me, her idea even, and this had to be done. "If you want to repay me, then you better get well. It's the only thing I ask. You have to be better the next time I see you." I don't think I've ever given anyone a longer hug in my life.

# 9

## SANTA FE

Those final days in LA, which were like so many of my days there, revolved around my finale with benzos. I wanted it to be just the Neurontin, if anything, that I got out of my system with the doctor's help, as there was some evidence online that Lyme patients actually used the drug—not to mention Neurontin was often used in detox facilities to get people off other drugs and the consensus was that it was a simple wean—but the benzos were pure tenacious poison. I was now down to one-tenth of a Valium pill, hardly more than the smallest crumb a pill cutter could yield. Sometimes I'd lose the pill cutter and it was just me in the kitchen crying into a butter knife, wasting pill upon pill with inaccuracy, me just trembling before little peaks of white on a wood cutting board that should have been lovingly reserved for vegetables for a family salad. Instead, pills, always pills.

If there was one enduring image of the end of my LA days, it was this: me at the foot of my bed, curtains drawn tight, wondering *should or shouldn't I?* at a tiny crumb, incandescent white, as if white was always pure, as if a pill was always medicinal, as if health was always treatable, me sitting there, hours upon hours going by, *should I or shouldn't I?* sun up and sun down, eyes focused on the collection of powder at the tip of my finger, no bigger than an army ant, *should I or shouldn't I?* and the day ending like that, until the next day, and there I was again, yesterday and tomorrow, *should I or shouldn't I?*

In the bathroom that final day, washing my hair, I thought of another bathroom, many lives ago, in Germany, the scissors, my wrist.

*Should I or shouldn't I?*

I had followed the latter impulse and almost survived it.

I was close now. I had to be close.

So many people, people I didn't even know, believed in me. Daily I got encouraging messages from people who had donated online. *Are you in Santa Fe yet? Have your treatments begun? Remember we're rooting for you! Can't wait till you feel better!*

In the final days of August 2012, my parents very reluctantly dropped me off at Sydney's opulent home, a new money estate all of marble and gold—not one but two new Jaguars crouched in a parking lot the size of my parents' whole condo. Michelle was coming the following day, and they had both planned the whole journey—a sort of *Thelma & Louise* but with three women, and the demon being illness, the endgame being a doctor. Sydney's home was as lavish as I'd guessed, part *Dynasty*, part Hamptons. She made us feel very at home in her own way. I got a physical therapy session with her bodyworker at her local country club the next day, and it actually felt comforting.

When Michelle got there, Sydney convinced us to take one of her aerobics classes, during which we both just did basic motions, laughing our way through most of it. I was still weak, but not too weak to laugh my way through eighties dance hits as Sydney hollered orders into a headset and dozens of overweight housewives zealously stomped and jumping-jacked and box-stepped along. Sydney kept everything floral and baroque, fed us all sorts of gorgeous spreads and wines and breads and cheeses, and it almost felt like a forever holiday in a movie, her glamorous everyday world.

We got on the road a day later. The drive to Santa Fe from LA is about fourteen hours, so Sydney decided we should make one overnight stop at Prescott, Arizona, where her old high school friend had agreed to house us in her rather luxurious casita in the desert. Upon arrival, we ate chips and dips and lots of frozen desserts. We sat in front of a giant flat-screen TV that could count as a small movie theater screen and watched Michelle's award-winning documentary that I'd wanted to see since Yaddo days, all about plastic surgery and her plastic surgeon father. It was hard to remember this wasn't a vacation. I found myself sleeping well like I often did on vacation, when sleep was less a responsibility but an indulgence. I found myself even a little bit hopeful.

It was the first time that it occurred to me that wealthy people had such options, how with money you could actually take a stab at life, no matter how bad the odds. Bills were paid without me seeing them, all sorts of needs of mine were met without me asking, and I was able to believe this was a story that could end well.

The drives were lovely, scenic and serene, full of funny jokes and radio sing-alongs and occasional moments of Michelle filming us, asking us to go on about our mental states. Sydney mainly drove, though Michelle helped, too, at parts, and I bemoaned that I was still too ill to drive. But I mostly felt okay, aside from a few times when my heart would race, my breath would be short, and I'd begin to feel faint.

"Don't worry, darling, your job is to hang in there just a bit more, because we're almost there, to wellness!" Sydney would almost yodel in those moments.

And soon there it was: the Turquoise Trail, that beautiful flat expanse of violets and greens and blues and browns, the richest hues of a desert of one's dreams, the Land of Enchantment indeed. When we got to Santa Fe, Sydney had already reserved a room for us all at La Fonda Hotel—where Jacob and I used to have many an enchiladas-and-margaritas dinner—wanting a few days there before they handed me over to Daniel.

Sydney was a bit suspicious of Daniel's offer to take me in and wanted to figure him out first. And so Daniel, who was known for cooking multicourse meals every month for a nice array of locals, arranged a big dinner in our honor for the next day.

Sydney had me get my hair done before the dinner party, so I'd feel better. As I sat there in an old salon I used to drive by all the time, small-talking with the hairstylist about my bizarre journey, how I'd lived there and how I'd now returned, I saw my old curls and the big glossy black hair with all its abundance and sheen come back, and for a second I saw that I was indeed beautiful. I myself felt so much smaller, so much plainer than my hair. But in the bathroom after it was done, as I shook out my nearly waist-long curls, I slowly applied makeup again—the first time in over a year—with the inspiration my hair afforded me.

In some ways the powder and glosses just sat on me, like a casket heroine, my beauty a bit frightening in its artificiality, no glow coming from within but rather from brushes and sponges. But I would be beautiful again, I would care, and I would like it, I told myself.

Sydney and Michelle convinced me to wear a nice dress and we made it to Daniel's only a bit late. It was the first time I'd seen him in over a year and he looked much the same: eyes that always sparkled like a doll's in their deep greens and blues and browns, with maybe even a hint of violet. His skin was that particular gold of only the outdoorsiest boys. He had a boyish flip of brown hair and wore clothes that were slightly too tight, but showed off his strong woodworker's physique. He was a full six years younger than me, which made him younger than my little brother. It was almost too much to look at him, but I knew in another life, he was my visual dream boy. He held on to me for a few extra beats after we walked in, and whispered into my hair, first sniffing as if taking it in with all he had, that I looked *so beautiful*.

"I'm so sick, though," was all I could think to say in return, my voice a hoarse whisper.

"I know, but that's all about to change." His confidence attracted me like nothing else. It wasn't that I hadn't heard such promises from men before—every one of them with their deep desire to solve my problems, it seemed—but this time I seemed ready for those words, eager to meet his conviction halfway.

Dinner consisted of four courses and was a joy. Daniel and I were in orbit in the center, Michelle and Sydney secondary supporting cast, and everyone else a sort of tertiary cameo. Nothing that night could break the spell—I'd journeyed here, through so much hell, to dine at the house of a new potential boyfriend, the house that I was about to share with him.

The next day, Daniel decided to join us on a long, slow hike in Abiquiú, near Georgia O'Keeffe's Ghost Ranch, my favorite place in the Santa Fe area. I felt relief as we wound our way up Chimney Rock again, seeing its gorgeous layers of purple, orange, and cream, looking at the endless sky with its perfect procession of rapid clouds, the weird stylish loneliness of Ghost Ranch. Daniel held my hand all the way back to town in the backseat. Home, a sort of home, really.

Sydney and Michelle were also there for my first meeting with my new doctor, Dr. Canfield, who took one look at my paperwork from the other doctors and said—while on Michelle's camera—that I "unequivocally" had Lyme.

Upon hearing it so nakedly, upon having it corroborated, I found myself trying to second-guess him. It had been years of people not believing in me and my illness, and suddenly I couldn't accept it myself.

"Are you sure?" I asked.

He smiled a bit, almost amused. "Yes, I see this all the time. Your bloodwork is fairly easy to read."

I asked him again.

"Porochista, you have Lyme disease," he said.

Michelle and Sydney both shifted uncomfortably. "I think it's just new for her, hearing it from a doctor finally," Sydney offered.

My eyes were full of tears but I didn't want to cry on camera. I tried to ask anything else, but it kept coming to my lips over and over. "Please. Just tell me for sure. Is this definitely it?"

He was used to this, maybe. He put a hand on my shoulder and met my wet eyes. "Yes, Porochista, you have Lyme," he said. "And we're going to help you get better."

In that moment, I said yes—to the diagnosis, to the journey, and to a future.

---

Within days, I had moved into Daniel's place and was on a new vitamin and supplement regimen, with an oxygen tank at my disposal and a rigorous three-day-a-week IV schedule with Dr. Canfield. Soon thereafter, I was on the road to having confidence in my new strange life, where my full-time job was wellness.

For the next six months I lived in Santa Fe, four of them with Daniel as his girlfriend. It was an odd life, a waiting game, a rehabilitation, a strangely beautiful no-life that now I sometimes miss. A favorite passage of mine in Susan Sontag's *Illness As Metaphor* involves travel and tuberculosis. She says, "the TB sufferer was a dropout, a wanderer in endless search of a healthy place. Starting in the early nineteenth century TB became a new reason for exile, for a life that was mainly traveling . . . Keats was advised by doctors to move to Rome; Chopin tried the islands of the western Mediterranean; Robert Louis Stevenson chose a Pacific exile; D. H. Lawrence roamed over half the globe. The Romantics invented invalidism as a pretext for leisure, and for dismissing bourgeois obligations in order to live only for one's art." Never did I relate to that more than during this period of living in Santa Fe, when my only reason to live was living itself.

It was a struggle to give my days meaning, to give my life any shape or purpose. In the mornings, I would collect fresh eggs from the chickens in Daniel's backyard and bring them to him in the kitchen, where he'd whip up our breakfast. He'd go off to his woodshop to work, and I'd sit propped against a half-dozen pillows in the bedroom, head connected to an oxygen machine, which was helping me with "Herx" reactions—the very painful die-off that happens when you aggressively kill a disease.

I couldn't read or write—my brain felt many months away from regaining concentration—but I could do other things well: swallow food again, nap in bits, meditate even. I'd occasionally take little walks through Santa Fe side streets and sometimes even make it to a massage. There was little I could do but wait my body out.

I also found a great deal of pleasure in tending to my outward appearance, since my internal one was on such a different clock of its own. I began returning for blowouts for my hair and alternating them with manicures. Suddenly I was the girl asking for big hair—big Victoria's Secret model hair, Dallas debutante hair, I kept insisting. It wasn't a look I'd ever attempted before, but to look larger-than-life vital had an appeal to me. It seemed an exaggeration of health and it also seemed like a complete transformation. I wanted to be anything but a sick girl. My long nails, sparkling claws of new colors every few days, also added to this surreal dimension. I was trying to add a layer to myself while all the other layers repaired themselves. And it was a harmless way to spend the time, when I wasn't in the doctor's office all day, hooked to IV machines, breathing my way through the chemicals coursing through my veins.

Daniel watched in awe and a little bit of horror. He and I were clearly mismatched—he was so very young and wanted to live in Santa Fe forever, put off by the other lives I'd lived in LA and New York—and before we could grow too close together we began to grow apart. Neither he nor I had realized what a full-time job my healing would be.

Dr. Canfield saw me every few days, sessions that would cost sometimes $1,000, with me taking over a hundred pills a day—different combinations every couple hours—and an IV regimen of glutathione, Myers, C drips, peroxide, major autohemotherapy, and UV photoluminescence therapy, plus procaine neural injections. He eventually added the local apitherapist, a Hoshin bee guru by the name of Voyce Durling-Jones, to the mix. She helped lots of Lyme people in town. Even though I had never been stung by a bee in my life, I turned myself over to her bee sting therapy—live bees brought right to my skin, who'd die after releasing their venom. I soon became used to my body being brushed by hundreds of taps and dozens of full stings on my head, my morning breakfast involving her royal jelly, propolis, and pollen. She became a critical healer for me, the only mystic I trusted.

By Christmas Daniel and I were over, and someone else was

in my life: Cosmo, a poodle puppy I'd ordered from a breeder, almost on a whim. I had seen an identical dog—this little serene black poodle with a perfect shape and mellow disposition—with a patient at Dr. Canfield's office. The dynamics of the IV room became such that you knew immediately who to like and who to ignore, sometimes six hours going by with your arm on a needle, needing all the small talk of the fellow patients to be pitch-perfect. I knew the dog owner was to be avoided because my favorite patient, Faye, a ninety-year-old ex-dancer who was married to a Los Alamos physicist, and who was dying of some awful cancer, immediately shut her eyes tight when this other woman came into the room. Faye had this way of pretending to be deep in sleep if someone annoyed her, and I had tried to pick up this habit, but the dog was too cute to ignore. The woman mentioned it was a *Moyen poodle*, and as I searched on my phone I found there were only two breeders in the country.

I told her my misgivings, as I had only had retired rescue greyhounds and rescue salukis—I never wanted to support a breeder.

"Oh, nonsense!" Faye came to life when the woman was gone. "You are ill. You can't risk it with a rescue!"

She was right—she was always right, always with a quick answer for everything, so sharp and so wise.

"Well, one day," I muttered.

"One day?!" Faye exclaimed. "Today is the one day. Come on! Why would you wait? You need that dog. You loved it."

I knew Faye was right, so I went as far as leaving a message on the phone of one of those breeders, telling myself that it was nonbinding to inquire.

"Mark my words, you will have that dog!" Faye declared, and indeed Cosmo, the only black dog of a litter in Georgia, was shipped to me at eight weeks old only two months later.

I brought him to the IV room one day to show everyone, but especially to show Faye. I was gutted when I heard the news—Faye had died just a few days before. But I knew she knew I would have that dog—she knew before me.

Cosmo the puppy and I moved to a friend's compound up in the lovely Hyde Park region of Santa Fe. I felt ready to make plans to move back East—I knew that was where my work was, and Los Angeles now haunted me as a place of illness, desperation, and addiction. I was even sending out CVs for teaching jobs on the East Coast. For me, looking to the East again meant an opportunity to rebuild myself as a writer, the way I did it the first time as an eighteen-year-old Californian who was plunging into the ultimate transformation of her life. I thought about New York City all the time.

---

My second novel had sold to Bloomsbury in November 2012 after over two years of submissions, in the same week that Dr. Canfield had announced I was in remission from Lyme. I had forgotten what this felt like, but I was finally feeling like myself. I was infused with a very solid gratefulness to be alive.

When people ask me when I finally began to feel well, I think of that early March of 2013, when I went to rent a sixteen-foot Penske truck to drive myself and the poodle cross-country to New York. I had never done that before, but my newfound wellness gave me incredible courage. At first I asked friends to come along, but their schedules didn't match my timing. Then I asked my mother but she felt it was dangerous and thought if she didn't come that it would discourage me from doing it.

In the end, I realized it would have to be all me, with a whole suitcase full of supplements, my other possessions, and my dog strapped to a doggie seat belt in the front seat. The day

I left, it was snowing hard in Santa Fe and I only had minimal 4WD mountain-town snow experience. I backed out of the narrow hilltop at my friend's compound, making it past the gates, and paused to breathe. Santa Fe looked foreign suddenly, icy and unfriendly, not the sunshine-splashed land of sunsets and healing that had nursed me to vibrancy. I breathed in and out, as if in a studied meditation.

I remembered how, just down the road, in that beautiful Artist Road house of mine, Jacob had knelt naked before me, eyes full of tears, face flushed, asking me to marry him, in the house filled with boxes of wedding dresses to try on and unsent wedding invitations. That was my only memory of that other Santa Fe life. The rest of my Santa Fe memories were of oxygen, IVs, bees, blood pressure and heart rate monitors, and so many good pills that they eventually pushed any bad pills out of the picture. And when I looked forward to New York all over again, the picture was not in focus. I could see myself in cracked montages or blurry smeared stills, fuzzy episodes, someone so young, full of a misleading energy that seemed to point to life but was still so far from it.

I drove by my favorite New Mexican restaurant and picked up some blue corn enchiladas and Christmas chile to go, turned the music on to AC/DC's "Back in Black," and drove, a little too fast, out of town.

The next few days felt like pilgrimages to places I'd only leave behind—a visit to old Chicago friends, a couple who now had a child in Boulder; a stop in Omaha, Nebraska, a place I'd never been but went steak-shopping in; Iowa City, where writers from the Iowa Writers' Workshop were looking forward to meeting me at a gastropub; even a stop in Chicago to meet a man I'd fallen into an intense correspondence with via OkCupid. On St. Patrick's Day, Cosmo and I made our last stop before NYC:

Pittsburgh. The city was a drunken mess and I remember frat-boy types roaring at my door, ringing the doorbell throughout the night.

The final leg of the drive was sad for me. I had become used to the open road. I had made up rules for myself: only listen to local radio, only race truck drivers during daylight, never linger for gas late at night, try to stop only at nonchains, never have less than a quarter tank, give Cosmo extra walks when you get gas. I had done well. No day was too unbearable. I had come to enjoy the open American expanses, the adventure of unknown hotels, the dinners alone, the breakfasts on the go, the sing-alongs to radio oldies. I never felt alone, and I never felt at an end. When you are on the road like that, everything is a beginning.

And so it goes for recovery. You are rich in new starts.

We made it to New York after a week on the road, and I have still, to this day, never quite looked back.

# ON BEING A BAD SICK PERSON

I've never been good at being sick.

When I was a child, I dreaded sick days like I dreaded summer vacation. School was a place to get away from home and home was where the problems were. While other kids faked sick, I faked well so I could avoid a day spent with my mother and father, whose work schedules were far from set through most of my adolescence.

Then later I was bad at being sick in other ways.

With Lyme disease, certain diets are recommended. It's been long believed that sugar, dairy, and gluten can exacerbate Lyme symptoms. A paleo or ketogenic diet was often recommended; another diet called "Bulletproof" required I drink butter coffee for the first half of my day. I rebelled from all these at some point, being a carb addict for one thing. Weight loss always came too easy for me, so I tended to feel best when I had other people's unwanted pounds on me. For the first year of my Lyme treatment, my New Mexico doctor recommended I mostly eat raw vegetables and to this day nothing feels more like medicine than a green juice.

I ate bread. I ate sweets. I ate junk food and dairy. Fries with mayo have probably been my favorite food since I can remember. And I went further with toxins: I drank, I smoked, I even occasionally did drugs.

There was medicinal marijuana in my first year of treatment, endorsed by my doctor, and which I gleefully took until one night I got too stoned and swore it off. In my first months of living with Daniel in Santa Fe, I had no idea he had a side business as a marijuana grower, his entire basement floor hydroponic weed of all varieties. He supplied me with pot butter for months until I cut myself off.

I had been attracted to the illicit since I was a child; there are many photos of me posing like old starlets with a crayon cigarette dangling from my mouth. I dreamed of buying packs from corner stores when my parents were out of town as a teenager. And my first week of college I bought a carton, not just a pack, of Marlboro Reds. I was determined. I sat in front of my full-length mirror and practiced, one after another, hoping five a day would get me addicted. By two weeks, I was hooked. I'd take my pack with me to Manhattan and walk through the busy streets, with a lit cigarette in my hand, feeling like the true New Yorker of my dreams. My first novel owed so much to those five-minute smoking breaks that punctuated those long unending fellowship days of nothing but writing to make a living.

In college I did drugs, but I thought that was the end of that. In my mid-thirties a reckless boyfriend with a taste for coke brought it back to me. It didn't last too long as I was ill before and after I knew him and was sure this would be a phase. And I hesitated—I was supposed to be long past that, but I told myself a line or two couldn't be that much worse than a few coffees. Within months, I was dabbling. On one particular decadent New Year I spent it up all night with him, going wherever the coke was. The next day, I woke up in the afternoon, my heart racing all out of beat. "I'm dying," I told him. I wasn't. "This will kill me," I told him. It could have.

So I stopped. But even now I dream of it. I've often joked Iranians are built for toxins. Red meat, whiskey, cigarettes, these were the staples of my ninetysomething-year-old grandfather's diet. Every Iranian I knew indulged in things that were supposed to kill you. I suppose I believe we were built to endure them.

Another part of it is the thrill of the sick person making herself sicker. If you know a part of you is always dying, taking charge of that dying has a feeling of empowerment. My body goes against me often, so what if I put it through that myself? At times, in the composition of this book, I snuck a cigarette or two, a reward for some paragraphs or pages. I watch the smoke, I watch the ash, I watch my fingers curl around them in familiar ways, I take in the smell and the stench like the most familiar ambiance of my upbringing. This is me, I tell myself.

I've told people online to ignore me, as I post photos of nachos and beer and cigarettes. I'm not part of "health Twitter," I tweet often. People ask me for advice, and I tell them to look elsewhere. I hate to exercise, I hate to meditate, I don't enjoy eating well, I never get proper sleep. I am not a poster girl for wellness.

I am a sick girl. I know sickness. I live with it. In some ways, I keep myself sick.

# 10

## *NEW YORK*

We arrived in the evening, a weekend evening, when the lights of the city announced themselves with full glare and bustling arrogance. Suddenly there it was: all that I associated with youth and coming-of-age, the development of my personality, my authorhood, my role as the writer and girlfriend and best friend and neighbor and downtown girl. The honks, the smell of hot nuts and pretzels, the shouts and squeals and hisses and guffaws, the puddles and trash, the rats and roaches, and, more than anything, bumper-to-bumper yellow cabs. I was home.

My old roommate Julie, of my CVS ramen meal days, was in Asia and had agreed to let me stay for a month on her couch, or maybe even more when she returned, provided Cosmo got along with her cat, Mouse. I joked that the apartment was a Sarah Lawrence alum halfway house. I had so many memories there, and it made sense that it would be my haven upon return. My roommate who was renting the spare room was Tara, an old friend also from Sarah Lawrence, loud and confident, techy and Buddhist, a former architect gone songwriter, an Alabama-born Iranian, who used me to connect to her Persian roots.

My next few weeks in the city went along these same lines—in other words, lots of fun, an endless homecoming celebration. I saw old friends, I took Cosmo on long walks to various dog parks, I ate at all my old favorite haunts, I felt alive again, like nothing had happened these past five years since Ryan and I had left Brooklyn for what we assumed was the beginning of a short but stable academic life in Pennsylvania. So many lives had gone by, and yet suddenly at thirty-five I felt younger than ever. I felt gratitude and a deep sense of liberation.

I also quickly began to forget what it was like being sick, tucking the narrative deeper and deeper into the recesses of my memory. I had no use for it anymore. I promised myself I would not go there again.

In my first weeks back, my savings dwindled to a point where I would barely be able to pay rent in the coming months. So I got a job as a researcher and writer at Gary Null's headquarters on the Upper West Side. I only knew a bit about him—he was a New York personality, often on local radio stations preaching about organic produce and FDA corruption and all sorts of anticancer tips. He was an eighties health guru, a bit wacky, a bit New Age, a bit of a conspiracy theorist—but my life in New Mexico made it so that he seemed quite normal to me. It also appealed to me that he was an alternative health guy.

A large part of my treatment was owed to alternative therapies—in the end, Dr. Canfield's treatments worked because he was able to pair more conventional antibiotic therapy with all sorts of out-there therapies that you had to go off the grid for. Gary Null spoke of Lyme often and seemed to advocate for many of my therapies—from bee venom to his absolute favorite, ozone. He was said to have a nurse come in once a week to give his whole staff ozone pushes through their arm veins.

Gary was ancient but also had the aura of a man who would

live forever. He was both ethical and corrupt, a sort of millionaire who'd become so by shunning the ways of millionaires. The office had the weird eighties space motif you see in depictions of cults or Scientology centers. "We need people who are not only hard workers but invested in our message and here to contribute to the cause and spread the word," his lawyer, who interviewed me, said. "We want you to be one of us."

I was hired on the spot, and my duties entailed basically updating and compiling material for Gary's subsequent books. He had written many, dozens and dozens, on everything from women's sexual health to geriatric wellness to alternative healing to brain nourishment. He self-published or worked with imprints of obscure publishers. He didn't care how they sold or who read them—he just did his part and put it all out there.

I embarked on this new nine-to-five life with few qualms. I knew it would be more of a spring–summer job, as I had teaching lined up in area colleges for the fall. I got my ozone shots, I got free lunches from his grocery store next door, and mostly I dodged the man and his employees and surfed the internet all day. I was happy to have a job, but I held to my own old principles—the key with "shit jobs" was to never be the best or worst, just mediocre, so they barely notice you, much less promote or lay you off. You wanted to be so-so, so no one would ask more of you—and yet not so bad that they would need to replace you. You didn't want to be a threat or a burden, just a barely visible team player who came in and out, earned wages, and forgot about it all.

Meanwhile, I embraced the chance to rejoin the literary community, and I was back to readings, galas, meeting with my new publishers at Bloomsbury, and other literary events.

One day, I got an unusual email from someone whose name I recognized, a certain writer named Carl. The subject header read "dashing bastard sitting next to you at last night's film screening." I was confused and then I realized I'd warned the man in the seat next to me that my friend and I would likely be leaving early (it was a long movie and we were starving). In the end, he had left before us. In the email, he mentioned that he had received my email address from a mutual writer friend, that he knew my work, and that he wondered if I wanted to ever hang out with him.

The line that got me, that I could not quite believe I was reading, was this: you seem vibrantly, gorgeously alive.

It had been so long since someone had seen me like that. Had I ever even seen myself that way?

I also knew his line had another significance—I knew Carl's story, as it was well known in the literary world: his wife had died of an acute illness a couple years before. I knew he had a young child, and I knew he was an excellent writer. I knew that, prior to his wife's death, he had been seen as a sort of literary bad boy and then his image had changed. He now gave off "caretaker" which of course appealed to me most.

We met and soon began dating. We fell in love almost immediately. He was brilliant and romantic. Every day on my way to work, I had a totally original, meticulously crafted good morning text message. It amazed me, how just this alone had the power to transform my day.

Carl helped me when I moved from living with Tara to a one-bedroom apartment that I found in a beautiful old landmarked prewar building in the Mount Morris district of Harlem. He also helped me edit the galleys to my novel and decide on covers. We read each other's writing, discussed books, played off plotlines, and gossiped about other writers. We'd meet up later in the day at various restaurants when he had a sitter or talk on the phone when he didn't. Eventually Carl and I even took to meeting at seedy hotels—the justification was that my Harlem apartment was too far and his daughter was at his tiny one-bedroom. But really we did it because we were writers and we both wanted and needed adventure, and we liked the narrative. We'd both suffered so much in recent years, and we felt we'd be together to enjoy this wild story for the rest of our lives. We were alive. Vibrantly. Gorgeously. Or so it felt.

In the middle of the summer, well into our romance, I started to feel ill again for the first time since leaving New Mexico—in the Lyme way, a very particular feeling of illness: a hot flash, a sudden fever spike, dizziness and nausea, convulsive sensations. I had not considered I'd be relapsing from Lyme in the future. This began to grow acute one weekend, so much so that I wrote my doctor and he recommended going to the ER, if for nothing else but some fluids. I texted Carl and received no response. From the ER waiting room, I texted again—nothing. I called—nothing. I was starting to get worried when I texted again from the ICU and he still gave no response. I spent the night at the hospital, texting him every few hours, begging him to write back, check in with me, anything, I needed him.

This was my first Lyme relapse—I had been warned about them, but actually experiencing one made me feel vulnerable and scared. By the time I was discharged in the morning—with a recommendation to go back on antibiotics briefly—I still had not heard from Carl. Only after I updated him via text that I was out of the hospital did he surface.

*Thank goodness you are okay.* That was all his text said.

*Where were you?* I demanded, upset.

It took him a second to write back. *I was wrapped up in some things and just not looking at my phone.* But I knew this was a lie—Carl was always by his phone.

Eventually he cracked and told me: *Look, I'm sorry. It's just that—I can't do hospitals. I can't do illness again. Not after years of my wife's disease and what she went through. The truth was I saw where you were and I was too scared. I'm sorry I failed you but I'm scared I will always fail you in this one area. I hope not eventually but this is where I am right now.*

I didn't write him back, not that day, not until I had some distance from the episode. It appalled me, that this man I loved would be a man who not only could not take care of me, but would disappear when I fell ill. It was possible that I'd be ill on and off for the rest of my life. What would happen? Would he improve? Or would I have to forgive this, accept this fair excuse that had to do with his history?

But what of my history?

When I finally broke up with him, he took the news badly. He faded out of my life, and the next time I heard of him it was in the context of his marriage to another much more successful writer. He had often wished we could be seen as a "power couple" and I was relieved he had that with someone else who could support the weight of that dream. I wished them well in my head, from afar, so many lives away I could barely remember his voice.

I remembered long ago breakups had the power to shatter me, but this new me could not do that anymore. Self-preservation now felt a vital part of my character. No one could come in the way of that. In some ways, I had waited my whole life to feel this strong.

For a long time, years even, I was able to hold on to that.

# ON LOVE LOST & FOUND

In the period between Carl and my last partner, I've counted nine partners. In just four years. It's a lot for me.

They all had two things in common: they were the wrong partners and our relationships had everything to do with their relationship to my illness.

There was Trace, a former-hacker-turned-artist, with the rich father. Trace, who had met me online, volunteering himself to accompany me to a Lyme documentary, who had come into my life specifically as someone who would take care of me. Trace, who said he loved me *for* not *in spite of* my illness, who claimed he'd helped heal a past girlfriend of her cancer. Trace, who brought me expensive juices and made me paleo pizza. And Trace, who was so rough with me in bed, every single time, that I could barely understand my own body for days after. Until I couldn't take it anymore.

There was Julian, the journalist who loved booze and drugs, who thought my illness wasn't such a big deal, who resented how much I planned and organized things for fear of physical setbacks. Julian, who ignored the idea of physical anything by numbing himself so much that life was all intoxication or the morning after, always an altered state. For a while Julian's hedonism seemed a way to forget my sick body—a way to kill my sick body so much I could forget it, rather. Julian, who loved to talk books and music till the sun rose; Julian, who had said he wanted to marry me on our first date. And Julian, who would just leave the room if I complained of any ailments for more than ten seconds.

Then there was Harley, who I still think impregnated me. I'll never know for sure but, based on my description, my gynecologist thought an early miscarriage might have been what that otherworldly period was about. Harley, the novelist, who had a young girlfriend back home—the girlfriend I eventually had to contact. I don't think I was the only one who experienced Harley this way. Harley, who never knew me as a sick person but definitely knew me as an addict, our connection was so potent and self-destructive and brief. But in the one month before that possible miscarriage, another reason that makes me think I was pregnant is that I felt better in my body than I had my whole life. Everything seemed glowing inside and out of me, my self whole and rounder and softer. (I've had high testosterone my whole life—and some think this has resulted in me being less-bed-ridden than some with my same Lyme numbers, testosterone known to be an activating hormone—but my body's ecosystem blossoms to peace, euphoria, and stability whenever I'm on female hormones like the pill.) When I flushed the toilet of that blood, I thought, Harley, I could never thank you for that. Harley, you could never know what we never had.

There were others: the graphic designer. The fashion photographer. The techbro.

There were women too—the famous writer, the academic, the scientist. (I tend not to dwell on bisexuality in these pages, but I've identified as queer since the mid-'90s. Because I am afforded heterosexual privilege in dating men so often, I tend not to rush to mark that box. Perhaps it's also because I feel overwhelmed by all my marginal identifiers. But I question that omission; to leave that out would be disengenuous too.)

All through this book, all through my life, I have loved many people. Too many people, you could say.

So many presented themselves to me as I got well and they all faded as I got worse.

At some point I questioned why I had for much of my life leaped from one person to another, with no end in sight. I'm not sure my conclusions are good, but I can tell you when the body feels out of place it will cling to anything that looks like life. Cities. Homes. People. Lovers.

Love is the only good way many of us know how to feel alive.

And the ghost I so often was wanted badly to feel real. And the characters in this section can at least tell you I existed. They might not have thought of me much, but they can tell you I was real. Sometimes too real.

# 11

## EVERYWHERE ELSE & AWAY

In Leipzig, the month before my health collapse, I remember I felt the healthiest of my life. I was suddenly sleeping so well, thinking so clearly, meditating, devoting so much time to health and wellness. And then I crumbled. Sometimes I wondered if it weren't those weeks before of goodness that were what really saved me, that healed me in the end, the body going back to recent muscle memory of high performance. Or another way to look at it: the body had given its all, devoted itself to all matter of excellence knowing it was at its end. Who can understand the machinery of our bodies totally? The logic seems always somewhere outside of us while in us.

I thought of that in the fall of 2015 and winter of 2016, which depended largely on my very healthy summer of 2015 for recovery. My recent past was where all my odds lay. If I had a chance it was because before some huge disaster, my system would take care of itself, as if it knew what was about to come.

The publication of my second novel in 2014 allowed me to travel the world in 2015—to London for book business, then to book festivals in Australia and Indonesia and eventually a book project in Israel. Indonesia in particular was a highlight. The Ubud Writers & Readers Festival in Bali was unforgettable. Going to Indonesia meant a lot to me, because it was my first time in a Muslim country since I left Iran in my early life. I was sponsored by the US Embassy to do satellite programs in Jakarta and Semarang, and in a bit over a week, I got to know diverse parts of Indonesia rather well. It turned out—to my surprise even—that it was one of my favorite places on earth. And that included the Muslim aspect—I found myself looking forward to prayer calls, trying to recite back, recording them daily, proud that the sounds of my culture were indeed so beautiful.

Our fancy Ubud resort hotel in what they called the monkey jungle was indeed full of monkeys and the most gorgeous lush forest, and beautiful breakfasts of *nasi goreng* and fresh coconut water and star fruit. The beauty at every step was astounding—the flora and fauna were breathtaking, but even the variety of prints on people's clothing, the sounds of gamelan everywhere, the smiles and kindness of people, it all blew me away. The college town Semarang had a sort of beauty to me as well, with all its motorcyclists—sometimes a four-person family all atop the motorcycle together—and winding roads. It was a dream.

I didn't seem able to shake the jet lag, however. In Indonesia I blamed my sleeplessness on humidity, but once back in New York I could not even come close to catching up on sleep. Something was off for sure.

I tried to blame it on the pressures of a difficult semester and the news—the Paris attacks had just occurred. Also my editor who had purchased this very book left publishing quite abruptly. The world felt unconquerably unstable. But something also felt wrong within me, and I could not figure it out.

I felt like my entire body was statically charged; everything felt electric in a menacing way. When I'd lie down, my body would buzz—I could not seem to tune out the activity in my own blood. My head throbbed one side at a time, headaches that felt unlike any headaches I'd had before. At times my hands and feet would go numb, other times it felt like my face was going to collapse in numbness. I started having the same swallowing problems. I couldn't open jars. I had trouble turning the knob on the door. In the shower, chunks of hair were falling out. Weird rashes were appearing on my face and limbs.

Was this a Lyme relapse? It had to be, and a bad one. I called my doctor in New Mexico. He pointed out that in the past six months I had crossed the international date line four times, with many internal domestic flights in between. That was a significant stressor.

I tried to ignore it. After all, at this stage I barely drank, I no longer smoked, I was back to working out, and I was supposed to be on the up.

But it got worse. I started doing funny things in my sleep. One morning I woke up in my bathtub, luckily dry. Another morning, Fresh Direct delivered box after box of baby food. I told them there had been a mistake but they insisted it was mine. I checked my account and indeed in the middle of a night that week, I had purchased $100 of baby food—little jars of every variety. Who could even begin to wonder why.

*Oh no, not again*, a voice inside me pleaded. *Nope, not again*, a more defiant voice insisted.

Finally, a couple weeks later, one very rainy night in November as I was driving home from teaching, I realized in the middle of the highway that I no longer knew where I was. I could not recognize the road, the names of things, where I had been, where I was going. I thought surely I was having a stroke. I got off the freeway and my old student assistant got on the phone and directed me to an urgent care where they immediately directed me to the hospital ER.

They did a CT scan and it came out normal.

"Well, this is good, I'm sure it's just still jet lag," I told the neurologist there.

But she was not convinced. "I wouldn't think so and I'd really check this out," she said. "Just because you're not being admitted doesn't mean there's nothing wrong, just that there is nothing we can do. But something is off. And you need to look into it."

I nodded. "Well, I mean, I have Lyme, so maybe. I don't want to get all wrapped up and take it too seriously again, you know."

She looked at me like I was crazy for being so casual. "Lyme is serious stuff. Why would you not take it seriously?" And that was the first time a physician in an ER told me that and took me seriously, not the other way around. I marveled at her statement for a moment, before I had my answer.

And then I realized I sounded like my own quack doctor. "Because I just don't want to go down that road," I said. "Not again."

# EPILOGUE

And then the car accident.

And then this book, The Book I Sold. The Book I Sold was a story of triumph, of how a woman dove into the depths of addiction and illness and got well. She got herself better. She made it. The Book I Sold might even imply you can do it too. Or anyone can. Who knows. The Book I Sold was never written past a barebones proposal.

Instead of writing The Book I Sold, in the winter of 2016 I fell into the darkest depression since 2011. That I survived to write this is the greatest miracle of all. That I managed to put words on these pages—in between my lengthy weekly sessions at NYU Medical's Concussion Center where I still get vestibular, cognitive neuropsych, and physical therapies—is shocking to me.

This book *is*, it turns out, a miracle book, because it wrote its own ending, insisted on its own ending. It didn't believe in my bows, my full circles, my pretty arcs, my character development.

It reminded me that illness will always be with you as long as life is with you. And tragedy will be with you too.

My car accident nearly killed me in ways I don't entirely

understand because it was nearly impossible to separate the Lyme from the concussion symptoms. And it was at this stage that I realized, thanks to my doctor in New Mexico, that I had one of the CDC-reported cases of Lyme because of my bands, those confusing barcode-looking codes that hold the secret to what's wrong with me. There was no arguing with the severity of my case; unlike the vast majority of Lyme cases in the country, mine did not need to be "believed." It qualified for medical belief.

And so when my doctor tested me again this winter, and I came out as positive as I was years ago, it was as if no treatment had ever taken place.

I had again become very, very sick very, very fast.

For a while I could barely rise out of my bed without a panic attack—*dysautonomia*, my doctor called it, as they realized my blood pressure dropped to alarming lows as my heart rate skyrocketed. I became someone who drinks salt water all day to keep my blood pressure in check, to keep from tipping over, falling, losing my head.

The landscape of my sleep became dramatically altered terrain, so that I only fell asleep around 6 a.m., without fail, and often I'd be dizzy in my sleep, and other times I would wake up from convulsions so violent they would throw me out of bed.

I lost twenty pounds in two months.

Not a day went by that I didn't cry violently for hours.

Not a day went by where I didn't know where I was for hours.

Not a day went by without those killer headaches.

Not a day went by when I didn't go back to thoughts of suicide, real ones, bright and hot.

I could not tell you how I walked the dog, how I ate, how anything happened. People came in and out and visited me, but everyone saw the same sight. I was locked in for months—me and my dog in an embrace, trembling on my red couch, which used to look like a pretty voluptuous tomato to me and now just looked like a bloody swollen organ.

My imagination became ugly. Twice I had those auditory hallucinations. Once it was the sound of the seashore on the phone with my friend Laura. The other time it was the sound of metallic wings flapping—like the illustration on the cover of my novel *Last Illusion*. It's a sound I almost knew.

And again every thought circled back to suicide. I was unable to imagine a way out because I could not read or write. Even my imagination was disabled.

There were pockets when I felt sane.

When I chastised my parents for again leaving me, always disappearing when illness or accidents happened to me, their unbearable fear always when I'm down.

When I spoke to certain friends.

When I got friends to take away the bee farm a friend bought for me, because another Lyme friend suggested I sting myself in the absence of my bee venom guru, and it was a disaster, my fingers unable to pinch the bees, the old honeybees dying in vain, their buzzing insanity-inducing in those endless insomniac nights, my soul feeling only all the more cursed with the psychic burden.

When I got myself, wheelchair and all, to the airport and to New Mexico to see my doctor for a week at the end of winter break, right before classes. He did every IV treatment but this time because of the concussion he would lose me at times—I would panic, couldn't tolerate things, I felt out of my head. I could not explain to him how I felt. My bee venom therapist, Voyce, felt it too. She in fact got into a similar car accident my last day in town and I felt it was because of my energy.

*All around me, things break*, I thought, *the most dear things.*

When I decided to take up my department chair's offer on

taking over my classes, because I am in no shape to teach much less walk through a campus.

When on Christmas Eve, which I spent alone, because absolutely everyone is gone, I agreed to speak to a woman who has been trying me for months, a very sick woman in San Francisco who is convinced she has Lyme, who is suicidal and desperate for guidance, a fellow writer. I talked to her and tried to fill her with all the hope I had now lost, the hope of The Book That I Sold, the hope that was essential, and it reminded me of me a bit.

When on my birthday, January 17, I was sitting, gaze locked to Twitter—alone again, friends wondering what they should do with me, but I had no preferences yet, still in that dark fog of injury and illness—me, marveling at the balloon icon graphics that appear on one's birthday on the site. Somehow even that electronic robot gesture of goodwill meant something. And I looked and I smiled and thought I should go to bed now, when suddenly there it went again: that heart racing, the pounding of the adrenals like bass, blood whipping around in me like a blender, head pulsing, body going to cold and there I was fading. I tried to get up and go to the hall and talk to my neighbor who was sitting on the steps.

When this time I knew better than to call 911 or let someone else do it, because the paramedics always take longer to get to Harlem, and I got in a cab, and in the cab, I nearly lost it.

"This is it, I'm gonna die on my birthday," I gasped, grabbing on to the seat in front of me as if we are on a roller coaster.

"When's your birthday?" the cabdriver, a Long Island guy, asked.

"It's now—today," I said, wheezing.

"Oh, happy birthday," he said, looking at my sorry state from his mirror. "Hey, you okay?"

"No," I nearly wailed. "That's why I need to go to the hospital. But quickly. I'm dying!"

He frowned at me. "Come on now. Hang in there. What's happening? You want to call someone?"

I thought about this. Call someone. Who did I call anymore but friends? *Someone else*, I thought. My mother. It made sense. *My birthday, my mother.*

I called my mother and immediately, like vomit, out came a chain of frantic Farsi, prayers and pleas all mixed up, my body going into convulsions, me holding on, the eyes of the cabdriver on me again and again.

My mother kept telling me I was going to be okay, she promised me, I was going to be okay.

I heard myself respond in an agitated Farsi and my mind went to the Paris attacks. I cut off my own rambling and I turned to the driver frantically, "Excuse me, sir, just so you know, it's Farsi. I'm Iranian, but not one of the bad people, please don't be worried by my language."

He looked surprised. "I grew up near Great Neck with you Persians."

I nodded. "Just so you know though . . ." Though I don't know who I was saying this to, what that disclaimer was for, what was happening to me.

A couple times I screamed and both my mother and the cabdriver told me to hold on, almost there, gonna be okay.

And this was one of those episodes where I felt sane because I was sane—it was my heart rate and blood pressure that were off, something was wrong with me for sure, this the ER doctors knew.

At Mount Sinai, I was immediately on a stretcher and nurses were all over me. Nobody knew what to do with me—is this cardiac, is this neurological, is this psychological, is this infectious disease?

EKGs and more imaging and more nurses and more vitals and more doctors through the night. I tossed and turned and occasionally I called a nurse over, asking that they just hold my hand. At some points, I would beg for oxygen. Friends text, parents call. I don't want to talk to any of them. The only person I contact while there is my landlord to make sure he is checking up on Cosmo, because he told my neighbor he would if I was ever ill.

*I'll be home soon enough*, I typed into my phone. I gazed into the fluorescent neon of the hospital, both clean and filthy, a place I felt comfortable in, after so many years of so many of them, all over the place. The beeps and the buzzing, the rapid sound of heels, the screams and the cries, the occasional expressions of joy—this was a home of sorts, I had to and hated to admit. I tried to kill the hours any way I could, and the only thing that came to me was the one thing that has managed to help a few times before: I imagined. I imagined myself many decades into the future, black hair now partially silver, body worn, face wrinkled, glasses. I imagined myself doing all sorts of things, this old lady of my dreams, hiking into those amber mountains of Santa Fe, another dog maybe at my side, squinting into the sun. I imagined myself in a sensible silk black dress, stepping out of a cab in Midtown Manhattan. I imagined myself swimming too far out at my favorite beach in Zuma, up the road from Malibu. I imagined myself back in Indonesia, falling asleep to the sounds of Muslim prayer echoing off the walls of the cities. And I imagined myself writing, reading, fingertips at a keyboard, eyes glued to a page, dazing off within the folds of a story, remembering the lines of a poem, purple pen to pages and pages of edits. And I imagined myself further holding a book, this book, This Book that was not The Book I Sold, holding it against my chest, if only to feel my heart beat against it. The story didn't end as I imagined so many times: in the end I would make it.

# ACKNOWLEDGMENTS

I've been writing seriously my whole life—since I learned to speak I would tell stories—but this third book of mine felt nearly impossible from start to finish and everything in between. Part of the difficulty in writing *Sick* was that I was quite sick during the making of it, and continue to be. (As I input final edits now, I am doing them between a lengthy hospital visit; even when the ink was barely dry, my failing body kept trying to rewrite this, to make sure it never ended.) You are reading the middle of the story, I suspect, but I'm not sure where or when it will all end, so one might as well tell it now.

The other challenging part was undoing so much of it to get to my story. At one point there was a lot of cultural criticism, historical anecdotes, facts and figures everywhere. Ultimately, I put fancy writer ego aside and stripped the book of all that and left it to be my story alone. I had to face that to tell this story honestly, I could not lean on anything but myself. While they may no longer be in the text, they are forever the gods of this book to me: Audre Lorde, Susan Sontag, Lucy Grealy, Virginia Woolf, and Anatole Broyard. Their illness narratives were my church. The words of Lucille Clifton also lifted me as I wrote as they have lifted me my whole life. "Won't you celebrate with me" is the real epigraph of this whole book, but I didn't want to drag her beauty into my mess.

I must thank my integrative medicine doctor and LLMD Russell Canfield of 360 Medicine in Santa Fe, New Mexico. He got me back on my feet in 2012 and led me to my Hoshin teacher and bee venom therapist Voyce Durling-Jones—to whom this book is dedicated. It is no exaggeration to say this team saved my life, and I continue to see them several times a year for treatment. Huge thanks also to Melissa Gzaskow and Sudarshan Ahben—it's a special place where even registered nurses and phlebotomists feel like family.

Dr. Harris of Malibu, California, was my first Lyme doctor, and I also have seen Dr. Horowitz in Upstate New York. It was nice to feel heard and believed—they had truly seen it all with this awful disease. I wish I could remember the name of the infectious disease specialist at Cedars-Sinai who thought I had Lyme before anyone else because he is a real hero of this book too.

I also want to thank the NYU Concussion Center and its team and staff—neurologists, neuropsychiatrists, physical therapists, vestibular therapists—for gluing me back together after my severe concussion in late 2015 through much of 2016. I thank them also for believing in Lyme and understanding how to manage a severe concussion that turns into a Lyme relapse.

I had a list of friends I contacted pretty regularly during the worst of my Lyme relapses: Benjamin Moser, Alexander Chee, Cal Morgan, Cassie Jones Morgan, Esmé Weijun Wang, Suleika Jaouad, Matthew Thomas, Laura van den Berg, Lauren Groff, Elliott Holt, Kris Jansma, Alyssa Harad, Ken and Aviva Baumann, Laura Coxson, Nathan Larson, Amanda Fortini,

Calli Ray, Sini Anderson, Shanna Mahin, Jennifer Pashley, Candice Tang, Deb Olin Unferth, Marie-Helene Bertino, Nam Le, Jon Durbin, Danny Rafinejad, Will Chancellor, Victoria Redel, Danzy Senna, Matthew White, Brooke Geahan, Candelora Versace, Christie Davis, Marie Mutsuki Mockett, Marcus Chang, Mark Lotto, Hillary Frey, Ana Cecilia Alvarez, Ales Kot, Brett Baldridge, Mensah Demary, Mike Scalise, Miranda Wonder, Melissa Ximena, Sahra Motalebi, Jaclyn Hodes, Ana Saldamando, Emma Forrest, Neil Gaiman, and Alanna Gabin. These people did all sorts of things: checked on me, cooked for me, did my dishes, walked my dog, went with me to doctor's appointments, sent me care packages and gift cards, held my arm as I walked on the street with my cane, stayed on the phone with me for hours and hours as I cried and cried with little hope in sight.

Special thanks to Mitch McCabe and Sarah Sleeper and Darcy Cosper, who pushed me to the first stage of wellness in 2012. Thank you everyone who donated to my GoFundMe—that plus my NEA paid for my first year of treatments in full!

Biggest thanks of all to Esmé and Suleika, the best 2/3 of our holy sick girl trinity, for saving my life over and over and over.

Thank you to my beloved readers Danzy Senna, Laura van den Berg, Matthew Thomas, Brett Baldridge, Cal Morgan, and Carvell Wallace.

I want to thank those living writers all around me who have written about illness so beautifully: Karrie Higgins, Sonya Huber, Alice Anderson, Sarah Manguso, Keah Brown, Alice Wong, Esmé and Suleika, of course, and so many more. Without these disability warriors and chronic illness activists there would be no me writing about this at all.

Thanks to my students and colleagues at all the places I

taught during the writing of this book: Fordham, Bard, Sarah Lawrence, Wesleyan, Columbia, VCFA, and Stonecoast—all families of sorts.

Thank you to the NEA, Ucross Foundation, Vermont Studio Center, the PEN Emergency Fund, Shuttleworth Foundation (thank you, Jesse von Doom, for my flash grant!), City of Asylum, and the MENA program at Northwestern University for allowing me funding and physical space to write. Thanks to The Wing (and Book Club!) for a place to also write in peace. Thanks also to Kundiman and Asian American Writers Workshop families for truly being a family to me.

It's a weird feeling to thank dead friends, but I feel their spirits all over this book: Cebrin, Samer, Mark, Maggie, Karin, Travis, Ellen, Rayya. All of your ends were untimely in every way. I think of you all often and question why I am here and you are not.

Thanks to all my friends and neighbors in my Harlem neighborhood: Tito, Rufus, Brahim, Mariama, Sarma, Leon. So many ambulances, so many ERs, so many errands, and so many dog walks. Thank you for making me realize time and time again I was not so alone as I thought I was.

Thank you to the whole Harper Perennial and Harper Wave families at HarperCollins. I lost two excellent editors and an amazing publicist during the making of this book (they are alive! they just got promoted at different publishing houses!), but one nice way to think of it is that I had many different teams at many different stages of this project. It truly took a village, or ten! All of their input was indispensable. Thank you, Cal Morgan, for acquiring this book, and thank you to Sarah Murphy for editing it, and a huge thank you to Laura Brown for taking it on at the last minute and doing a bit of everything in making it all run so spectacularly. Thank you to Amy Baker for seeing this through from start to finish, in so many ways—you are a legend for a reason. Thank you to my publicist Lily Lopate who also got passed on so much second-hand and patiently rose to challenges. And of course, a huge thanks to my agent Seth Fishman, who for close to a decade now has stuck with me through so many dead ends, rocky roads, rough air, near misses, and even the occasional smooth sailing!

Thanks to Carvell for more wisdom, empathy, and love than I ever could have imagined. Thanks to my parents for enduring their child's endless nightmares of body and mind—I would not want to be my own parent. And of course much love and biscuits to my superstar dog, Cosmo: support animal, dogson, poodlebro, critter. He's always been there in a way that allows me to pretend he has another choice.

And finally thank you to the many, many, many sick people around the world who reached out to me and urged me to write this book. And especially the women of color who told me they needed this narrative for more reasons than one. I never ever planned on writing a book like this; truly, this was all for you.

# ABOUT THE AUTHOR

Author photograph by Sylvia Roskoff

POROCHISTA KHAKPOUR was born in Tehran in 1978 and raised in the Los Angeles area.

She is the author of the novels *The Last Illusion* (Bloomsbury, 2014)—a 2014 "Best Book of the Year" according to NPR, *Kirkus Reviews*, *BuzzFeed*, *PopMatters*, *Electric Literature*, and more—and *Sons and Other Flammable Objects* (Grove, 2007)—the 2007 California Book Award winner in "First Fiction," a *Chicago Tribune*'s "Fall's Best," and a *New York Times* "Editors' Choice."

Her other writing has appeared in the *New York Times*, *Los Angeles Times*, *Wall Street Journal*, *Al Jazeera America*, *Bookforum*, *Slate*, *Salon*, *Spin*, CNN, *Daily Beast*, *Elle*, *Conjunctions*, *American Short Fiction*, *Gulf Coast*, and many other publications around the world. She's had fellowships from the National Endowment for the Arts, University of Leipzig (Picador Guest Professorship), Yaddo, Ucross, and Northwestern University's Academy for Alternative Journalism, among others. Currently she is a guest faculty member at the Vermont College of Fine Arts and the University of Southern Maine MFA programs as well as a contributing editor at *The Evergreen Review* and *The Offing.*

*Sick*—a "Most Anticipated Book of 2018" according to the *Boston Globe, BuzzFeed, HuffPost, Nylon, The Rumpus, The Millions, Bitch, Bustle, Autostraddle,* and more—is her first memoir.

In 2019, a collection of her essays, *Brown Album,* will be published by Vintage. Pantheon will be publishing her next novel, *Tehrangeles, in* 2020.

She lives in New York City's Harlem with the love of her life, a standard poodle named Cosmo.